WITHDRAWN
UTSA Libraries

ENVIRONMENTAL RESEARCH ADVANCES SERIES

TUNDRAS: VEGETATION, WILDLIFE AND CLIMATE TRENDS

ENVIRONMENTAL RESEARCH ADVANCES SERIES

Environmental Research Advances. Volume 1
Harold J. Benson (Editor)
2008. ISBN: 978-1-60456-314-6

Handbook on Environmental Quality
Evan K. Drury and Tylor S. Pridgen (Editors)
2009. ISBN: 978-1-60741-420-9

Estimating Future Recreational Demand
Peter T. Yao
2009. ISBN: 978-1-60692-472-3

Bioengineering for Pollution Prevention
Dianne Ahmann and John R. Dorgan
2009. ISBN: 978-1-60692-900-1

Bioengineering for Pollution Prevention
Dianne Ahmann and John R. Dorgan
2009. ISBN: 978-1-60876-574-4 (Online Book)

Sorbents: Properties, Materials and Applications
Thomas P. Willis (Editor)
2009. ISBN: 978-1-60741-851-1

Environmental Cost Management
Randi Taylor Mancuso (Editor)
2009. ISBN: 978-1-60741-815-3

Biological and Environmental Applications of Gas Discharge Plasmas
Graciela Brelles-Mariño (Editor)
2010. ISBN: 978-1-60741-945-7

River Sediments
Greig Ramsey and Seoras McHugh (Editors)
2010. ISBN: 978-1-60741-437-7

Biogeography
Mihails Gailis and Stefans Kalniòð (Editors)
2010. ISBN: 978-1-60741-494-0

A True Tale of Science and Discovery
Lawrence A. Curtis
2010. ISBN: 978-1-60876-595-9

Amazon Basin: Plant Life, Wildlife and Environment
Nicolas Rojás and Rafael Prieto (Editors)
2010. ISBN: 978-1-60741-463-6

Handbook of Environmental Research
Aurel Edelstein and Dagmar Bär (Editors)
2010. ISBN: 978-1-60741-492-6

Behavioral and Chemical Ecology
Wen Zhang and Hong Liu (Editors)
2010 ISBN: 978-1-60741-099-7

Advances in Environmental Modeling and Measurements
Dragutin T. Mihailović and Branislava Lalić (Editors)
2010. ISBN: 978-1-60876-599-7

Tundras: Vegetation, Wildlife and Climate Trends
Beltran Gutierrez and Cristos Pena (Editors)
2010. ISBN: 978-1-60876-588-1

ENVIRONMENTAL RESEARCH ADVANCES SERIES

TUNDRAS: VEGETATION, WILDLIFE AND CLIMATE TRENDS

BELTRAN GUTIERREZ
AND
CRISTOS PENA
EDITORS

Nova Science Publishers, Inc.
New York

Copyright © 2010 by Nova Science Publishers, Inc.

All rights reserved. No part of this book may be reproduced, stored in a retrieval system or transmitted in any form or by any means: electronic, electrostatic, magnetic, tape, mechanical photocopying, recording or otherwise without the written permission of the Publisher.

For permission to use material from this book please contact us:
Telephone 631-231-7269; Fax 631-231-8175
Web Site: http://www.novapublishers.com

NOTICE TO THE READER

The Publisher has taken reasonable care in the preparation of this book, but makes no expressed or implied warranty of any kind and assumes no responsibility for any errors or omissions. No liability is assumed for incidental or consequential damages in connection with or arising out of information contained in this book. The Publisher shall not be liable for any special, consequential, or exemplary damages resulting, in whole or in part, from the readers' use of, or reliance upon, this material.

Independent verification should be sought for any data, advice or recommendations contained in this book. In addition, no responsibility is assumed by the publisher for any injury and/or damage to persons or property arising from any methods, products, instructions, ideas or otherwise contained in this publication.

This publication is designed to provide accurate and authoritative information with regard to the subject matter covered herein. It is sold with the clear understanding that the Publisher is not engaged in rendering legal or any other professional services. If legal or any other expert assistance is required, the services of a competent person should be sought. FROM A DECLARATION OF PARTICIPANTS JOINTLY ADOPTED BY A COMMITTEE OF THE AMERICAN BAR ASSOCIATION AND A COMMITTEE OF PUBLISHERS.

LIBRARY OF CONGRESS CATALOGING-IN-PUBLICATION DATA

Tundras : vegetation, wildlife, and climate trends / editors, Beltran Gutierrez and Cristos Pena.
   p. cm.
Includes index.
ISBN 978-1-60876-588-1 (hardcover)
1. Tundra ecology. I. Gutierrez, Beltran. II. Pena, Cristos.
QH541.5.T8T87 2009
577.5'86--dc22
                              2009043034

Published by Nova Science Publishers, Inc. ✦ *New York*

CONTENTS

Preface		ix
Chapter 1	The Change of Tundra Biota at Yamal Peninsula (The North of the Western Siberia, Russia) in Connection with Anthropogenic and Climatic Shifts *M. G. Golovatin, L. M. Morozova, S. N. Ektova and S. P. Paskhalny*	1
Chapter 2	The Reciprocal Relationships between High Latitude Climate Changes and the Ecology of Terrestrial Microbiota: Emerging Theories, Models, and Empirical Evidence, Especially Related to Global Warming *O. Roger Anderson*	47
Chapter 3	Soil and Plant Characteristics in the Alpine Tundra (NW Italy) *Michele Freppaz, Gianluca Filippa, Angelo Caimi, Giorgio Buffa and Ermanno Zanini*	81
Chapter 4	Modern Climate Trends and Possible Changing of Arctic Coastal Zone (Russian Sector) *Sergey Nikiforov, Vladimir Byshev, Ogorodov Stanislav and Putans Victoria*	111

Chapter 5	Alpine Meadow on the Tibetan Plateau was a CO_2 Sink in Peak Growing Season Revealed by Keeling Plot Method *Xiaoyong Cui, Hongchao Tan, Yibo Wu, Jing Wu, Yangong Du, Yongcui Deng and Yanhong Tang*	**131**
Chapter 6	Tundras and Climate Change: A Mammalian Perspective *Věra Řičánková, Jan Robovský and Petr Pokorný*	**151**
Chapter 7	Carbon Deposition on the Forests of Some Treeline Ecotones of the Ural Federal District *A. Usoltsev Vladimir*	**161**
Chapter 8	Genetic Diversity and Population Structure of Alpine Plants Endemic to Qinghai-Tibetan Plateau, with Implications for Conservation under Global Warming *Yupeng Geng, John Cram and Yang Zhong*	**175**
Index		**193**

PREFACE

Tundra ecosystems are seriously affected by global climate change. Understanding tundra history and postglacial development may enhance the ability of biologists to anticipate biotic responses to current environmental changes. In this book, the authors analyze changes which have occurred in a vegetative cover and aboveground fauna of vertebrates at Yamal peninsula, one of the greatest plains on the globe. The authors also evaluate pedogenetic processes, soil nutrient status and plant distribution along an elevation gradient in the alpine tundra in the western Italian Alps. In addition, treeline ecotone is a belt of transition from forest vegetation to a nonforest one, which allow the monitoring of climate change. In this book, carbon deposition on the forests of two treeline ecotones is studied. Some of the current emerging theories, models and recent empirical evidence for the dynamics of these reciprocal interactions between climate and terrestrial microbial communities are also reviewed, with particular attention to biogeochemical and ecological perspectives.

Chapter 1 - The directed changes of biota usually explain by two most obvious reasons – anthropogenous influence and climatic changes. At the last decades a climate change is the subject of wide speculation. The strongest trend of change of air temperature in high-altitude areas is observed on periphery of Northern Atlantic and Northern Eurasia (Briffa, Jones, 1993), including in the north of Western Siberia (Rubinshteyn, Polozova, 1966; Shiyatov, Mazepa, 1995). At the same time the Yamal peninsula differs from other areas of Arctic regions presence of the powerful anthropogenous influence connected, on the one hand, with enormous overgrazing of domestic reindeers, with another – with management of hydrocarbonic fields. It is possible to assume as climate warming and anthropogenous influence should be reflected on biota of the region. In the work we have analysed changes occurred in a vegetative cover and terrestrial fauna of vertebrates at Yamal peninsula. Thus paid attention only to the cores or focal points that though simplifies an overall picture but, on the other hand, does by its more distinct.

Chapter 2 - High-latitude, moss-rich tundra communities (e.g., *Sphagnum* and *Hylocomium* spp.) are circumpolar in distribution, including conifer forests and tundra ecosystems that occupy millions of square kilometers. The sheer geographic scale of these high latitude biomes is sufficient to warrant scientific interest. However, it is becoming increasingly clear that major changes in high latitude climate patterns may have significant affects on the ecology of these communities. In turn, changes in the life histories, physiology, and productivity of the biota may also directly, or indirectly, influence local to global climate patterns; especially the balance of atmospheric carbon dioxide that is sequestered by primary production versus that released by respiratory activity – thus, potentially influencing global warming. Substantial attention has been given to aboveground biota, particularly the role of plants in this biotic-climatic reciprocal relationship, notably in relation to global warming and likely changes in annual mean temperature and precipitation patterns across vast geographic regimes at high latitudes. However, belowground processes also are likely to be substantially affected, especially the response of microbiota. Changes in the biology of terrestrial microbial communities may be directly affected by local meteorological factors, but also indirectly by effects of above- and belowground coupling. This coupling includes the effects of climate variables on plant physiology, especially the degree of primary productivity, release of organic compounds into the soil and their influences on the productivity and respiratory activity of associated belowground microbiota (e.g., bacteria, fungi and eukaryotic microbes, including protozoa). The major groups of protozoa include heterotrophic flagellates, naked amoebae (lacking a shell) and testate amoebae (enclosed in an organic or mineralized shell). With increasing evidence that the tundra permafrost is incurring prolonged seasonal warming and thawing to greater depths, there is an increased probability that associated microbial communities, that are normally more dormant during much of each annual climatic cycle, may become increasingly metabolically active. Given the enormous stores of plant-derived organic matter that have accumulated and remained frozen during millennia, there is substantial potential for enhanced terrestrial microbial respiration and significant release of atmospheric carbon dioxide. Nearly one-third of the global terrestrial carbon is stored in these high latitude environments. Currently, there is increasing interest in the complexities of the responses of terrestrial microbial communities to high-latitude climatic changes and the likelihood that they could have a significant effect on global warming through elevated respiratory activity.

Some of the current emerging theories, models, and recent empirical evidence for the dynamics of these reciprocal interactions between climate and terrestrial microbial communities are reviewed, with particular attention to biogeochemical and ecological perspectives.

Chapter 3 - The arctic tundra forms a circumpolar band between the Arctic Ocean and the polar ice caps to the north and the coniferous forests to the south. Smaller, but ecologically similar regions found above the tree line on high mountains are called alpine tundra. Here the soils are perennially cold and often snow-covered and the plants show a high degree of specialization in order to survive in such extreme conditions.

In this work, we aimed to evaluate pedogenetic processes, soil nutrient status and plant distribution along an elevation gradient in the alpine tundra in the western Italian Alps.

The study area for this investigation is located in North West Italy, close to the Monte Rosa Massif (4634 m asl). Soil profiles were dug and described at 5 sites along an elevation gradient from 2525 m asl to 2840 m asl, in the upper part of a glacial valley. Genetic horizons were identified and physical properties described following standard methodology. Bulk samples were taken from each major horizon, and smaller known-volume samples were taken for determination of bulk density. In the laboratory soil chemistry and particle size distribution were determined following standard methods. Data on the vegetation structure were collected close to each soil profile, covering a surface of 32 m^2; each sampling site has been further divided into 8 sub-areas of 4 m^2. The abundances of species were recorded as cover percentages.

Soil depth in the five study areas ranges from 10 to more than 50 cm. Soil profiles exhibit a fairly consistent horizonation. All of the profiles have a black to dark brown A horizon, in some cases underlying thin Oi/Oe horizons. Structure is usually weakly developed fine granular, except at lowest elevation, where is moderately developed subangolar blocky. Textures range from sandy to loamy sand. The major pedogenic processes responsible for soil formation in this environment are: a) organic matter accumulation and melanization; b) cryoturbation; c) erosion and deposition of material. Melanization appeared as a sharp contrast in darkness between the surface horizons and deeper parts of the solum. The accumulation of organic matter is encouraged by low temperatures that reduce rates of decomposition. The presence of irregular horizon boundaries within the profiles are evidence of cryoturbation, as these soils are frost affected, mainly in early winter and late spring, when the reduced snow cover may not be sufficient to insulate the soil from the air temperature. Further, soil erosion and deposition may occur under steep slopes, determining a loss and a subsequent redeposition of soil material and thereby playing a role in pedogenetic processes.

A total of 64 plant species were found at the sampling sites. As expected, the harsh alpine conditions influenced both the life form and the chorology of taxa. Hemicryptophytes (91.07%) were dominant, followed by chamaephytes (6.82%), geophytes (1.21%) and therophytes (0.91%). From the phytogeographical point of view, the species were mainly south european mountain (49.44%), artic alpine (29.74%), west alpine (7.81%) and alpine (6.82%) while the ones with wider range, as european or eurosibiric, were less frequent (6.19%).

Soils described in this area are surprisingly well developed. The organic carbon content of the surface horizons is quite high, and comparable with forested soils at lower elevation. From the top to bottom positions the synecology of plant populations depicted some clear gradients connected to changes in the most important landscape variables: decreasing in bryophyte cover, increasing in vascular plant cover, nitrogen content and in temperature.

Chapter 4 - The current climate variability allows establishing that natural processes are in line with the most important and observed global warming factors. The climate system in mid 70-ties of the previous Century entered an equilibrium which brought about a rapid global reallocation of the atmosphere. The concomitant alteration of the planetary atmosphere circulation brought about intensification of meridional warmth transfer from the Indian Ocean to the central areas of the Eurasian Continent and from the Pacific Ocean to Alaska. It was exactly in those areas that the large scale subsurface temperature anomalies in the last quarter of the XX-th century took place. The first decade of the XXI century revealed the features characteristic of the climatic system return to the state antecedent to the mid 70-ties of the previous century. Thermodynamic conditions of Arctic Ocean are defined by a number of factors, among which of greatest importance are:

- basin waters circulation and exchange with Atlantic and Pacific Oceans;
- heat flow balance on the Arctic basin surface and presence of energetically active zones with intensive ocean-atmosphere interaction and transformation of Atlantic waters;
- regional atmospheric circulation and connection thereof with planetary circulation.

Colligation of hydrogeological data provided by various Arctic expeditions in past century leads to the deduction of considerable warming (for 0.5-1.5C) of Atlantic waters in Arctic basin observed in last half of XX Century. Arctic warming inevitably affects all components of natural environment, including dynamic and morphology processes in coastal zone of Arctic seas. Simultaneously with climate warming the rising of sea level occurs. This also is a very important factor of future changes in coastal zone.

Decrease of area and existing time of ice sheet in Arctic Ocean seas and, as a consequence, increase of hydrodynamic activity in coastal zone will undoubtedly be one of the most important factors of changes for coastal zone dynamics in XXI century.

The most substantial changes in coastal zone dynamics and morphology will occur in Arctic seas with unconsolidated permafrost environments and intensive thermal erosion. The velocity of shores distraction is evidently been largely determined by deposits composition and iciness, both dependent on lithogenic environment and following cryogenic reconstitutions. Modern velocities of Russian Arctic shores thermal erosion vary widely from first meters to tens and more over per year in general. Rapid changes in environment can tilt a fragile balance and cause shore degradation. This is particularly important for Arctic shores which erode and step back nearly everywhere. Global warming aggravates the situation. Ice-free period extension contributes to increase in total near-shore wave energy. Ice-verge back off in summertime brings about the same effect of wind acceleration enlarging. The mentioned changes can have different influence on different shores. Thermal erosion will definitely increase. Thermal erosion, thermal denudation, deflation can bring about considerable economic loss and bring to naught profitability of raw material extraction.

Chapter 5 - Alpine meadow occupies about 37% of the total 2.5×10^6 km^2 of the Tibetan Plateau. Gas exchange between this ecosystem and atmosphere may have a large impact on greenhouse gases budget of the whole plateau. It is still under debate whether alpine meadow is a sink or source of atmospheric CO_2 in growing season.

Air samples were taken along vertical profiles in the boundary layer above three communities: *Kobresia humilis* alpine meadow, *K. tibetica* swamp meadow, and *Potentilla fruticosa* shrub meadow, in growing season in 2007. Relative contribution of gross photosynthesis (GP) of the vegetation and gross respiration (R, including soil and plant respiration) to net ecosystem exchange of CO_2 was calculated by constructing Keeling plots.

GP was greater than R for *K. humilis* alpine meadow and *P. fruticosa* shrub meadow in clear days, indicating carbon sinks of these communities. The ratio of GP/R was higher in the former community than in the latter one, suggesting stronger sink for the *K. humilis* alpine meadow. GP/R decreased to near 1 when it was cloudy in these two ecosystems. During showers, the points deviated largely from the lines in Keeling plots. However, progressive transition between sink and source was discernable in these plots.

GP was roughly similar or slightly lower than R for the *K. tibetica* swamp meadow even in clear days. The swamp meadow accumulated deep peat layer during its development and may continue sequestrate C in the soil. Large quantity of exotic dissolved and particulate organic carbon transported from surrounding communities and from herd excreta in the swamp meadow were supposed to result in high gross respiration in the this ecosystem.

The results implied that the alpine meadow ecosystem on the Tibetan Plateau was the sink of atmospheric CO_2 in growing season.

Chapter 6 - Tundra ecosystems are seriously affected by global climate change. Understanding tundra history and postglacial development may enhance the ability of biologists to anticipate biotic responses to current environmental changes. Tundra-steppe was dominant vegetation type during Last Glacial and supported rich mammalian (mega)fauna.

Climate change represents critical factor responsible for the extinction of Glacial tundra-steppe fauna. Relatively constant climate in Central Asia allowed for a preservation of the tundra–steppe fauna in Altai-Sayan refugium. Arctic tundra could be considered rather a product of climate changes then anthropogenic activities and possibly represents a Holocene novelty, at least from a mammalian perspective. There is no evidence of extinction of arctic tundra species during Holocene. If the magnitude of current global warming anticipated by recent scenario remains unchanged, we may well expect the effect of climatic change analogous to that during "climatic optimum" of the Middle Holocene.

Chapter 7 - Carbon deposition on the forests of two treeline ecotones is studied. The first of them is on the altitudinal gradient of the western slope of Tylaiskii Kamen Mountain (the western part of Konzhakovskii-Tylaiskii-Serebryanskii Mountain system, 59^0 30' N, 59^0 00' E), namely on the belt of the upper tree-limit rising between 864 and 960 m above see level, during the last 100 years. The second one is a zonal ecotone near the lower Pur river (67^0 N, 78^0 E) as the transition belt between closed flood-lands forests and island-like forests on the tundra watershed varying from tens of meters to 2-3 kilometers depending of relief peculiarities and flood-lands width. On the first treeline ecotone the 5-6 times decreasing of the carbon pool in *Picea* biomass between altitudinal levels of 864 and 960 m a.s.l. was recognized. On the second ecotone at the age of 45 years and similar densities (1300-1700 trees per ha) the carbon pool in *Larix* aboveground biomass and needle biomass on the flood-lands are correspondingly 7,0 and 2,4 times more than on the watershed. In senescent forests this difference is some more, correspondingly 10 and 3 times. Annual carbon deposition differs 5 times by these two sites.

Chapter 8 - The Qinghai-Tibetan Plateau is one of the most important centers of biodiversity for alpine species in the world and is among the areas that are most sensitive to global warming. Knowledge about population genetics is essential for understanding the dispersal ability and evolutionary potential of alpine species in a warming world. In this chapter, we review the genetic diversity and population structure of 19 alpine plant species endemic to the Qinghai-Tibetan Plateau. Generally, the population genetic variation can varygreatly among different species and the endangered species have much lower levels of genetic diversity than the co-occurring common species. Although a few species showed increased levels of genetic diversity along altitude, we dectected no significiant correlation between diversity and altitude in most species. In addition, the isolation-by-distance model cannot explain the spatial genetic structure in most alpine species that have been investigated, which may partially due to the discontinous distribution of alpine species shaped by complex geomorphology in Qinghai-Tibetan Plateau. The implications of these results for the conservation of alpine plants during global warming are discussed.

In: Tundras: Vegetation, Wildlife…
Editors: B. Gutierrez et al. pp. 1-46

ISBN: 978-1-60876-588-1
© 2010 Nova Science Publishers, Inc.

Chapter 1

THE CHANGE OF TUNDRA BIOTA AT YAMAL PENINSULA (THE NORTH OF THE WESTERN SIBERIA, RUSSIA) IN CONNECTION WITH ANTHROPOGENIC AND CLIMATIC SHIFTS

M. G. Golovatin[*], L. M. Morozova, S. N. Ektova and S. P. Paskhalny

Institute of Plant and Animal Ecology,
Ural branch of Russian Academy of Sciences, Ekaterinburg, Russia.

INTRODUCTION

The directed changes of biota usually explain by two most obvious reasons – anthropogenous influence and climatic changes. At the last decades a climate change is the subject of wide speculation. The strongest trend of change of air temperature in high-altitude areas is observed on periphery of Northern Atlantic and Northern Eurasia (Briffa, Jones, 1993), including in the north of Western Siberia (Rubinshteyn, Polozova, 1966; Shiyatov, Mazepa, 1995). At the same time the Yamal peninsula differs from other areas of Arctic regions presence of the powerful anthropogenous influence connected, on the one hand, with enormous overgrazing of domestic reindeers, with another – with management of hydrocarbonic fields. It is possible to assume as climate warming and

---

[*] Corresponding author: E-mail: golovatin@ipae.uran.ru

anthropogenous influence should be reflected on biota of the region. In the work we have analysed changes occurred in a vegetative cover and terrestrial fauna of vertebrates at Yamal peninsula. Thus paid attention only to the cores or focal points that though simplifies an overall picture but, on the other hand, does by its more distinct.

METODICAL PECULIARITY

Study Area

The Yamal peninsula is located in the north of the Western-Siberian lowland – one of the greatest plains on globe (Figure 1). Accordingly it is one of few flat areas of Arctic regions. Absolute heights from 1-5 m at coast hollow rise to an axial part of peninsula to 25-50 m. The greatest high-rise mark – 91 m. The surface is strongly enough dismembered by valleys of the rivers, streams and lake hollows.

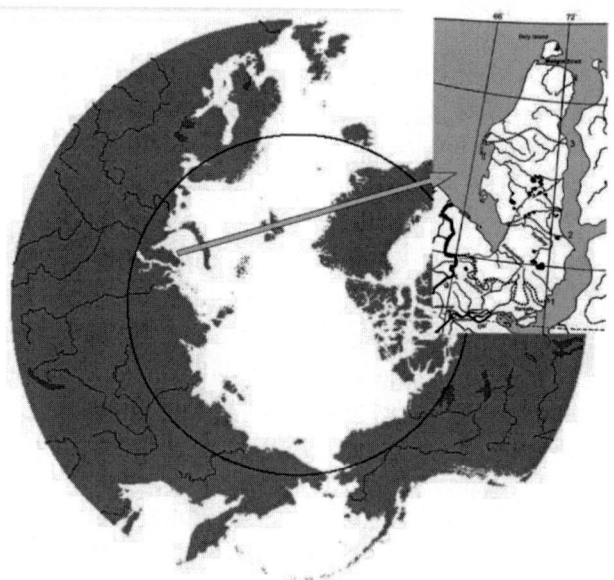

Figure 1. Geographic location of Yamal peninsula and zonal stracture of vegetation cover. Borders of zone and subzone (by Atlas of YNAR, 2004): 1 – forest-tundra and southern Subarctic tundra; 2 – southern and northern Subarctic tundra; 3 – northern Subarctic tundra and arctic tundra.

Yamal entirely lies in a permafrost zone. Unlike the majority of the Arctic areas the basic composing breeds are friable quaternary sediments, it is strong sandy and sated with ice. The big extent of peninsula territory from the south to the north (750 km) and smoothness of relief cause distinctly expressed of zonality of a climate and a vegetative cover. About 80% of territory make Subarctic tundra (to 71°N). This subzone subdivided into strips of Shrub or Southern tundra (to 69°N) and Typical or Northern tundra (69-71°N) that occupy accordingly 50 and 30% of territory. Typologically and physiognomicaly their vegetation is very similar, the basic difference concerns a density and height of shrub layer. In southern tundra shrubs (dwarf birch *Betula nana*, willows *Salix* ssp., an alder *Duschekia fruticosa*) inhabit on watersheds, in northern – the alder is absent and dwarf birch and willows form the rarefied undersized stands on slopes. Bogs widespread both in river flood-lands and on watersheds also occupy 22-24 % of territory. In the south of peninsula in river valleys there are large forests, individual trees go out on watersheds.

To the north 71°N the subzone of the Arctic tundra (20% of territory of Yamal) is located. Typical zone communities here are dwarf shrub-lichen-moss tundra at tops and slopes of watersheds and grass-moss tundra in relief falls. Bogs make 16% of territory. Their feature is spotty grass-moss cover and insignificant capacity of peat. In the north of peninsula shrubs are absent even in river valleys.

Administratively Yamal is included into the Yamalsky district of Yamal-Nenets autonomous region (YNAR) in Russia.

Methods

We compare results of own long-term researches (1980-2008) of vegetative cover and population of vertebrate animals to the published materials of last years. For an estimation of large-scale changes of a vegetative cover used V.N. Andreev's data (1934). Special works under the census of terrestrial animals in considered area began to spend only with 1970. Before researches were limited general observations of survey character on travels on the region. During such travel on eyes visible species (active and loud, large, numerous, lodging near settlements or living in open habitats), first of all, come across. Therefore, discussing change of numbers and distribution of species, we speak only about those from them which it was necessary to notice. We take present cases of registration of secretive, not numerous and small animals into consideration only at comparison with those publications of last years in which the species composition is precisely enough reflected. At comparisons statistical significance of distinctions defined by Styudent test.

In the researches we combined works on stationary plots (the size 10-50 km^2) with observations on excursions (the pedestrian, boat, cross-country vehicle). In total on Yamal 70 plots (by a total area 1750 km^2) have been surveyed: on Northern – 21, on Middle – 29, on Southern Yamal – 20 (Figure 2). The period of field works had for June-August.

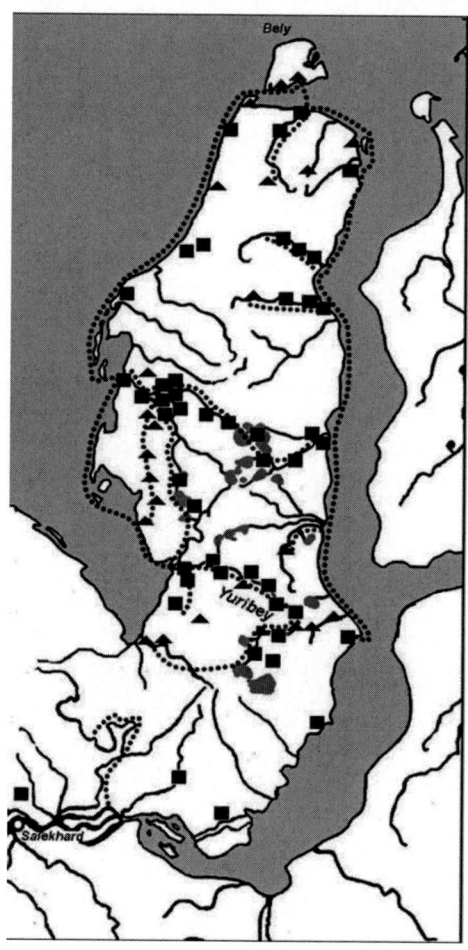

Figure 2. Map of author's investigations on the Yamal (1 – field station with investigations during 2 and above seasons; 2 – field station during 1 season; 3 – motor boat and half-tracked vehicle trips).

Vegetation studying realized by routeing investigation and ecological profiling with use of a method of the standard geobotanical description. On ecology-topographical profiles from the river flood-land or lakes to tops of watershed hills described all vegetative communities. For this purpose on plots in the size 10×10 m fixed the general projective covering, a covering on sublayers and synusia (shrubs, dwarf shrubs, grasses, mosses, lichens), revealed full species diversity of vascular plants, mosses and lichens, height of grass and a thickness of a moss-lichen cover. The abundance of species of vascular plants, mosses and lichens was estimated by the Drude scale or in percent. Total number of geobotanical descriptions for research years was about 1850.

The stock of aboveground phytomass was estimated by a method of hay crops. Sampling was realized on registration plots in the size 25×25 cm in number 5-10 pieces. Herbaceous plants and dwarf shrubs was cut off at level of border of green and brown parts of mosses. Lichen-moss turf was cut out by knife, in the absence of it (on the trampled sites) remains of mosses and lichens were collected in packages. In laboratory conditions samples were assorted on fractions (grasses, sedges, herbs, dwarf shrubs, lichens, mosses). The samples have been dried up to an air-dry condition and weighed.

In a northern tundra subzone of peninsula in 1993-1995 experiment for the purpose of monitoring of vegetative cover recovery process out of reindeer grazing influence has been put (Magomedova, Morozova, 1997). In three types of the most widespread vegetative communities plots in the size 5×5 m have been allocated and fenced from access of reindeer (Foto 1). Later 11 years (in 2006) we have visited three remained plots and surveyed vegetation on them. On two plots the samples on the phytomass supply was taken in the way specified above: by 5 samples inside and outside of each plots.

Estimation of a share of the sandy exposures formed under the reindeer grazing influence was spent on plots in the size 80-180 km^2. Plots were selected in different parts of peninsula on the basis of images from Google Earth Program. In total 65 plots have been used: 14 – in southern tundra subzone, 29 – northern subarctic tundra and 22 – in arctic tundra. Natural sandy exposures in the form of river spits and the dried lakes did not take into account. The percent of sandy exposures from the land area (without reservoirs) was considered.

Foto 1. Experimental plot for study of vegetation recovery without reindeer grazing (Western Yamal, Northern Subarctic Tundra).

At birds censuses we preferred absolute census on large plots (Robertson, Scoglund, 1985) by method of intensive mapping of territories (Gudina, 1999) with the subsequent recheck of the data. The size of plots was for passerines and waders 1-15 km^2, for other groups – to 50 km^2. Level of rodents abundance defined in marks using the complex data of censuses in the different ways: catching by traps (usually 50 traps, established through 5м on the lines located in different biotopes), by a dog, visual calculation of wintering nests and moving rodents on routes.

The Yamal peninsula favorably differs from other high-altitude areas of Russia that here on the meteorological station Salekhard (WMO #233300, 66°31' N, 66°36' E, 35 m above-sea level) there is the most long number of tool meteorological observations (125 years, since 1883). Its comparison with the data collected on younger meteorological stations of the north of Western Siberia shows very high degree of similarity in dynamics of characteristics (Shiyatov, Mazepa, 1995). It is logical to believe that for Far North biota existing in the conditions of heat deficiency the climatic changes occurring during the snowless period when there is a vegetation of plants and reproduction of animals have more great value. Therefore we have considered the data about monthly average temperature of air in spring (May) and summer (June – August) periods.

Besides for this region there is a possibility of use of indirect estimations of climate changes for longer intervals of time by results of dendroclimatologycal analysis considering variability of a year radial gain of trees (year rings of wood). In the chapter results of reconstruction of average summer air temperature in a southern part of Yamal peninsula for 4309 years (Hantemirov, 2000) are used. Correspondence by sign test between a course of the reconstructed air temperatures and instrumental data reaches 82 % and on the average is 76% (Mazepa, 1999). It points to a split-hair accuracy of dendroclimatologycal reconstruction of summer air temperatures.

CLIMATIC CHANGES

During the period of instrumental supervision on a meteorological station Salekhard is observed a trend of increase in monthly average temperatures of spring and summer months (Figure 3). However, the period beginning is suited on the end of a cold climatic epoch which is well looked through on dendroclimatic reconstruction (Figure 4). If to take that into consideration and to exclude a piece to 1908 within last 100 years of essential increase in temperature it was not observed, only insignificant lifting in May and July, the temperature of August even has decreased (Table 1).

Table 1. Average air temperature of spring and summer months on the meteorological station Salekhard in the first and second half of 100-year period.

Period (years)	Average air temperature ± SD (° C)			
	May	June	July	August
1908 – 1958	−1,3 ± 2,7	8,6 ± 2,1	14,0 ± 2,0	11,6 ± 2,0
1959 – 2008	−1,1 ± 2,7	8,6 ± 2,3	14,6 ± 1,9	11,1 ± 1,6
T – test (p)	0,41 (0,69)	0,06 (0,95)	1,46 (0,15)	1,53 (0,13)

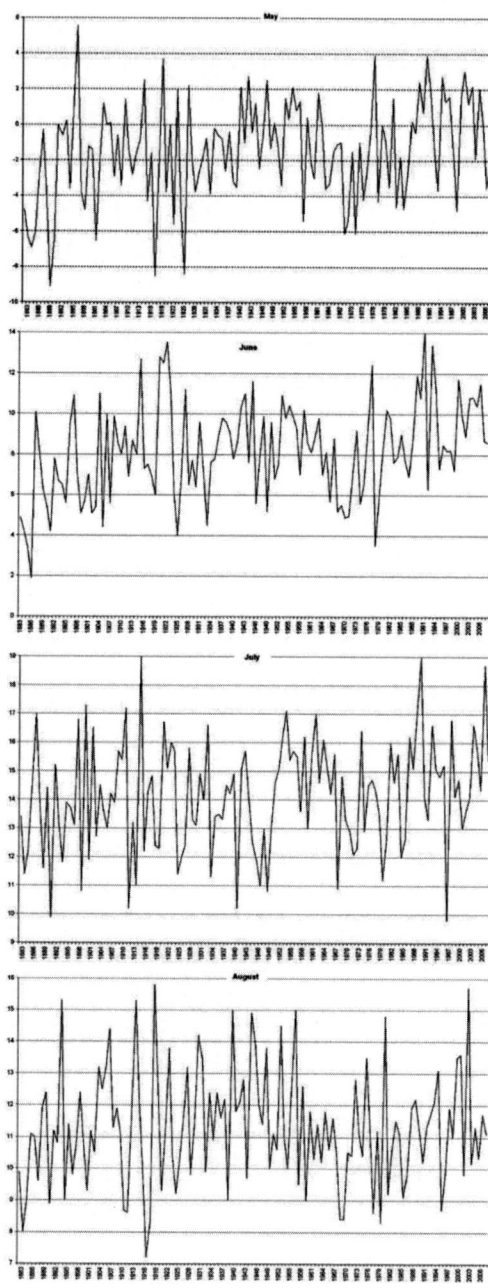

Figure 3. Dynamics of average monthly air temperature of spring-summer months at Salekhard meteorological station.

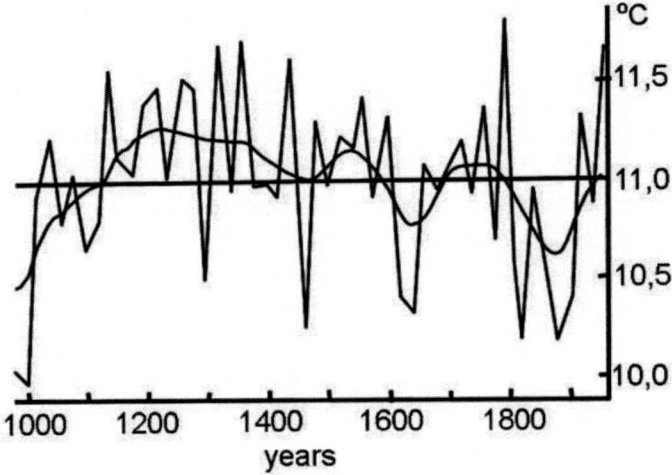

Figure 4. Reconstructed air average temperate (by 20-th years) of June-July for lasted thousand years for Salekhard (from Shyatov, Mazepa, 1995).

By consideration of deviations of temperature from many years' average for last 100 years a cyclicity in alternation of the periods with warm springs and summers is observed (Figure 5). Positive deviations of May temperature (early warm spring) were frequent in 1892-1897, 1904-1910, 1941-1959, 1987-2006. June was warmer usual in 1920-1923, 1953-1959, 1989-1994, 2000-2006. The long periods with high summer temperature (July) were in 1921-1924, 1952-1969, 1987-1998, 2004-2008. The most often warm autumn (August) happened in 1904-1909, 1931-1949, 1994-2003.

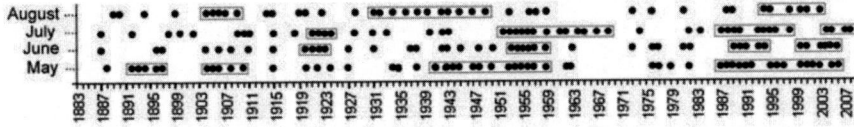

Figure 5. Years when average air temperature of spring and summer months was moved aside more than 0,5° C from average many years' temperature for 100 last years. Periods of frequent recurrence of such deviation is marked by contours.

THE HISTORY OF ANTHROPOGENOUS LANDSCAPE FORMATION

The history of anthropogenous landscapes formation on Yamal, as well as in the majority of other areas of Subarctic region, covers the little more than 100 years, except for solitary ancient human settlements in a southern part of area. Background change of initial landscapes under the influence of the traditional form of managing of the indigenous people – pasturable reindeer breeding become the most large-scale.

Up to the end of XIX century Yamal Nenets used domestic reindeers as a vehicle. Herds practically did not exceed 300 heads. In 1880 their neighbours – komi-izhemtcy which had already skills of large-herd outrun reindeer breeding, began to pass with their herds with western to an east side of Ural mountains. Nenets have estimated advantages of reindeer grazing by large herds and reindeer breeding became a basis of their life-support. Very soon (in the first decade of XX century) the number of reindeer on Yamal has reached 80 thousand (Tarasov, 1915). During the same time there were formed basic wandering routes which steadily remain since then. Their directions are mainly meridional.

Number of reindeers on Yamal and in YANR varied synchronously (Figure 6).

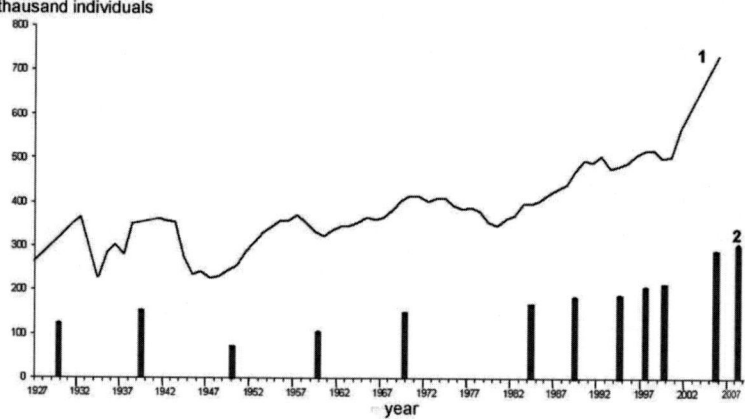

Figure 6. Number dynamics of domesticated reindeer in Yamalo-Nenets autonomous region (1) and on Yamal peninsula (2).

To the middle 1960th it remained at one level, in 1980th the steady tendency to growth was designated, especially sharp growth has occurred in the late 1990th at the expense of increase in a livestock at private owners after disorder of «collective» farmings of the Soviet period. Now in region the largest in the world livestock of domestic reindeers is concentrated – 730 thousand heads (40% of world numbers!). On Yamal about 300 thousand reindeers are grazing. Grazing is realized by large herds (about 3-12 thousand individuals) (Foto 2).

Foto 2. Grazing herd (Western Yamal, Northern Subarctic Tundra).

Average density of reindeers on peninsula is 2,3 individuals/km^2 (32-33 hectares of pastures on 1 reindeer), local – in tens times above. At existing system of grazing routes of movement of large herds can be crossed more than once. It is possible to choose some gradations of grazing loading on territory. Incredible loading on a vegetative cover is created when a herd in 10-12 thousand heads grazes on the same pastures twice for a season. Sites with such loads are located in convenient places for passage of herds. Similar loading is created by herds of numbers about 5-10 thousand heads grazing on the limited area and passing on the same pastures for a season to 3 times. Very high loading arise at unitary passage of herd to 8-12 thousand heads. Such loading is typical for most parts of peninsula territory. High loading is arised by unitary pass of herds about 4-8 thousand heads or 2,5-5 thousand heads twice for the season. Besides, small private herds migrate without plan on all pastures crossing ways of migration of large herds. The part of them remains on Yamal for wintering and terminates pastures already used in the summer. A practiced everywhere whirling of herds completely destroys vegetation and condenses top horizons of soils. There are not other more weak loadings on the peninsula.

As a result there was a unique situation for the tundra zone at which pasturable loadings have appeared exorbitant not on any limited local area but in enormous territory.

To 1920 on Yamal there were only small settlements mainly the fishermen camps attracting to a valley of Ob. The village Obdorsk (with 1938 – town Salekhard) was largest of them. In 1920-30th years at development of Nordic Sea Way and collectivisation a number of settlements and trading stations in a tundra zone has been based. Gradually there was a population growth. In the late 1950th – the beginning 1960th hydrocarbonic fields exploration has begun. In 1970th and especially in 1980th rates of an increase in population at the expense of immigration have sharply increased. In 1980-90th years organization of Bovanenkovo and Kharasavey gas fields has begun. In 1988 along Urals railway building to gas fields has started. The network of open-cast mines, service roads was generated, there were working settlements. Due to a lining of the car road in parallel the railway the southwest part of Yamal became accessible to motor transport that has increased recreational loading. Now the centres of technogenic influence are localised on the small areas of gas and oil fields prepared for exploitation and in a narrow strip along transport mains – pipelines, motor roads and railways.

TRANSFORMATION OF VEGETATION COVER

For a tundra zone the natural dynamics of a vegetative cover coupled with erosion-accumulative activity of water and wind and cryogenic processes is typical. As a result there are long cyclic local changes of vegetation. For example, in forest-tundra the cycle of change of spotty tundra through larch forest or shrub islet again to spotty tundra is about 350-500 years by some calculations (Kryuchkov, 1976). We did not consider similar natural cyclic dynamics of vegetation but speak only about scale changes of a vegetative cover.

Climatogenic dynamics of vegetation. Dendrochronologycal researches of the higher forest boundary in Polar Urals show synchronism of changes in structure and productivity of forest stands of forest-tundra communities in mountains and on plain (Shiyatov, Mazepa, 2007). According to these researches in the north of Western Siberia from the beginning VIII and till the end of XIII centuries there was an intensive expansion of wood vegetation which was expressed in essential increase in the areas of light forests and close forests. During a cold snap which occurred from the end XIII and up to the beginning of XX century forests receded. According to the data about a gain, structure, productivity and spatial distribution of forest-tundra tree stands the coldest was XIX century. As show results of research present warming is comparable with one in XII-XIII centuries but the tree vegetation yet has not reached the climatic caused limit of XIII century.

By results of archeobotanical analysis (Fedorova et al., 1998) in IX-XIV centuries in tundra at latitude 71°25' N the yerniks (dwarf birch *Betula nana*) were quite usual and formed shrubs. Now northern border such shrubs is almost on 100 km to the south at latitude 70°27' N. It testifies that in medieval warming the subarctic tundra subzone on Yamal has been considerably shifted to the north. Evidences of any scale changes of tundra vegetative cover connected with climatic cycles of smaller periodicity or last warming are absent.

Changes connected with technogenic influence. Active industrial development of Yamal, an event last 20 years, was accompanied by increase in technogenic loadings at a vegetative cover. However according to economic-vegetation map (1995) the area of technogenic disturbed grounds on Yamal in 1990th was only 0,5% from the total area. The general territory occupied with all deposits (exploitable now and perspective) and also planned transport linear constructions is about 2% of the peninsula area. I.e. now the changes of a vegetative cover connected with technogenic influence have local character and are insignificant on the area. At that on preparing deposits the area of completely disturbed grounds (inhabited settlements, industrial zones, roads, open-cast mines, mineral arenas, etc.) is nearby 1% of their territory (Table 2).

After forming of production infrastructure on deposits (pipelines, roads and other communications, settlements and industrial constructions) recovery of the disturbed sites begins. They overgrow with aboriginal species of grasses: the humidified and rehumidified sites by cotton-grasses, sedges and *Arctohpila fulva*, drainable sites and sandy mounds by graminifolius plants. At that the species diversity in secondary communities decreases (Figure 7). On our observations in territory of settlements the abundance of such species as *Tripleurospermum hookeri* Sch. Bip. and *Phippsia algida* (Soland.) R. Br. considerably increases.

Table 2. Disturbance of Kharasovey and Bovanenkovo gas field territories, % (Magomedova, Morozova, Ektova et al., 2006).

Degree of vegetative cover disturbance, %	Khrasovey	Bovanenkovo
90-100	0,2	1,5
50-90	1,3	2,4
10-50	11,0	46,7
<10	87,5	49,6

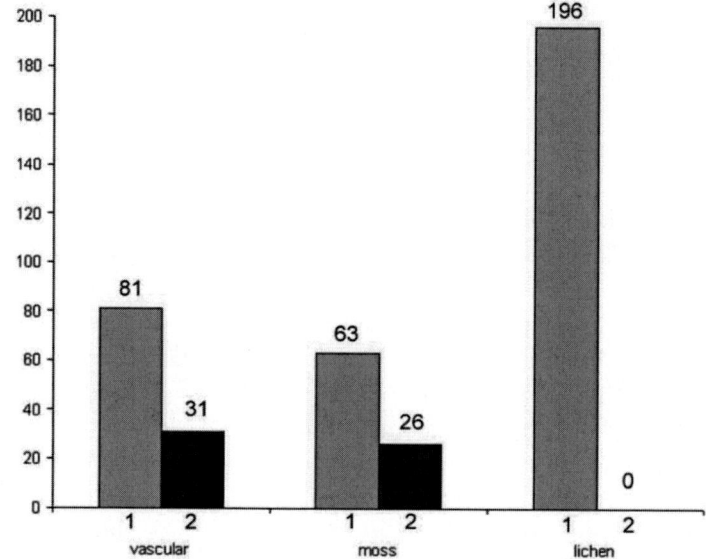

Figure 7. The species number of plant in native (1) and disturbed (2) communities on the territory of Kharasavey gas field.

In process of formation of a technogenic landscape there is a penetration on the north by carrying of some southern plants. For example, *Vicia cracca* L., earlier found only in forest-tundra of Yamal (Rebristaya, 2006), along constructed railway Obskaya-Bovanenkovo has penetrated in shrub tundra. In 2005 we have found out it on a railway embankment near the Erkuta-yakha River (68°13' N). Drift of "alien" species is promoted by existing practice of revegetation of disturbed grounds when species of plants are used atypical for tundra subzons. So on the Bovanenkovo gas field (in northern subarctic tundra) in 1989-1994 for aptitude for revegetation have been tested 10 boreal species of grasses typical for northern forest-tundra (Kotelina et al., 2000). One of them *Beckmannia borealis* (Tzvel.) Probat. we have found in 2005 on overgrowing sandy open-cast mines, roads, on landslides. In 2006 as a part of primary plant aggregation on the road slopes has been found typical carryed species – *Poa compressa* L. (Andreyashkina, 2008), the nearest which place of natural distribution is fixed on distance of 1100 km (Interactive Agroatlas, 2008).

Changes connected with reindeer grazing. Reindeer grazing covers all territory of peninsula entirely. As a result of an escalating press of pasturable loading the vegetative cover on Yamal has essentially changed.

It is considered that overgrazing leads to «greening» tundra when instead of lichens and mosses grasses is spreading (McKendrick et al, 1980; Olofsson et al., 2001). There is an opinion that thus even primary production of communities raises (Zimov et al., 1995, Olofsson, Oksanen, 2002). However on Yamal it does not occur. In comparison with 1930th supplies of grasses in tundra and on bogs (table 3) have decreased in 1,5-2 times. Grasses expand only on the small sites enriched by excrement – around Arctic fox holes and on long-term stands of reindeer breeders. On places with high and very high pasture load the grassy vegetation is foraged away to the height 3-5 cm. Often coefficient of biomass exception reaches 90 %. The tundra looks «shaved» as a fresh sheared lawn (Foto 3).

In tundra on sites with very high loadings the share of dwarf shrubs having for reindeers secondary food value has increased. Grazing have an effect on not only in a grassy cover but also on shrubs. General supplies of aboveground phytomass of willow stands have decreased: short willow stands – almost in 8 times, tall ones – in 2 times. The registration of number of gnawing parts of willow twigs showed that at moderate pasture loads the annual growth of branches was eaten on 52% of dwarf birchs and 67% of willows. Under very high loads shrub stands are oppressed much more strongly: shrubs are broken off and poorly foliaceous, the annual growth of lateral shoots decreases in 1,5 times, many branches is dry. In separate places where herds are passed along the same valley some times for season willows are presented as weak oppressed shrubs with a small number of the young shoots growing from roots.

Foto 3. Tundra under intensive pressure of reindeer grazing (Western Yamal, Northern Subarctic Tundra).

Table 3. Resources of phytomass (air-dry mass, ton /hectare) on Yamal in 1933 (by data of V.N. Andreev, 1934) and 1995-1997 years (by data of authors).

Groups	1933	1995-1997	
		lim	M±SD
Lichen and lichen-moss tundras			
Total	0,5-3,0	0,30-2,12	1,02±0,58
Sedge and grasses	0,2-0,7	0-0,31	0,17±0,10
Lichens	0,3-2,3	0,07-1,40	0,42±0,38
Dwarf shrubs	--	0,02-1,82	0,43±0,40
Grass- and dwarf shrub-moss tundras			
Total	1,3	0,6-1,47	1,02±0,27
Sedge and cotton-grasses	0,7	0,38-0,50	0,44±0,05
Grasses	0,2	0,19-0,25	0,21±0,02
Herbs	0,3	0-0,15	0,07±0,07
Dwarf shrubs	0,1	0,04-0,64	0,25±0,24
Lichens	до 0,05	0-0,16	0,06±0,06
Grass-moss bogs and boggy tundras			
Total	2,7-4,0	0,82-3,39	1,66±0,71
Sedge and cotton-	2,0-3,0	0,47-2,00	0,96±0,42

Groups	1933	1995-1997	
		lim	M±SD
grasses			
Grasses	0,3-0,4	0,35-1,39	0,70±0,29
Herbs	0,4-0,6	--	--
Flood plaine meadow			
Total	4,8	2,63-4,95	3,79±1,11
Short willow stands (0,3-0,5 m)			
Total	5,2	0,86-1,43	1,13±0,25
Leaf of willow	4,5	0,4-0,8	0,57±0,16
Sedge and cotton-grasses	--	0,12-0,34	0,20±0,09
Grasses	--	0,03-0,07	0,05±0,02
Herbs	0,7	0,05-0,19	0,11±0,06
Dwarf shrubs	--	0-0,05	0,03±0,03
Lichens	+	0,01-0,39	0,18±0,16
Tall willow stands (1-2 m)			
Total	3,6	1,09-2,92	2,03±0,75
Leaf of willow	2,2	0,61-1,80	1,13±0,50
Sedge and cotton-grasses	1,1	0,40-0,93	0,74±0,24
Herbs	0,3	0,08-0,19	0,15±0,05

The major factor of grazing influence on tundra vegetation is trampling. The fruticose fodder lichen species especially *Cladina* and *Cladonia* genus is the most sensitive to trampling. On Yamal in comparison with 1930th the lichen tundra area has been reduced in 3,5-4 times (Figure 8). The tundras are partially trampled down to barren ground and partially transformed in moss and dwarf shrub type of tundra communities with participation of lichens. The prevailing height of the lichen cover has decreased from 3-4 to 0,5-1,5 cm. At the beginning of the 20th century the resources of lichen mass reached to 6 ton/hectare (Andreev, 1934), at present time on most parts of lichen-moss tundra it hardly reaches 0,8 ton/hectares. For evident comparison of surface aspect of native and modern transformed lichen tundra we demonstrate photos of the tundra located at the same latitude: on Taz peninsula around the Yamburg gas field where reindeer grazing is limited and on Middle Yamal in the Basin of Yuribey River (Foto 4).

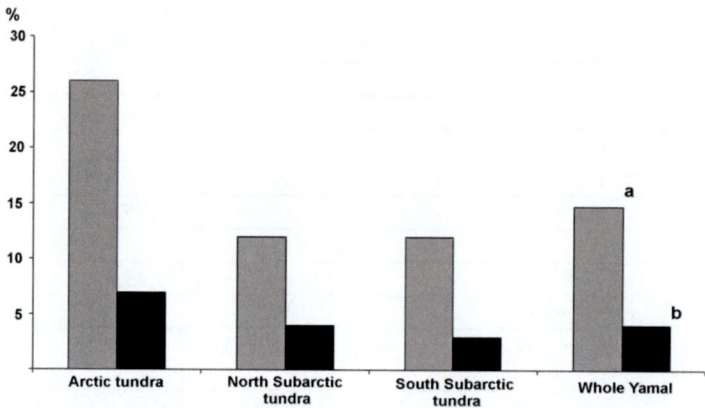

Figure 8. Changing of lichen tundra share at the vegetation cover of Yamal (% from covered by vegetation space) during 80 years: a – V.A. Andreev's data (1934), b – own authors data (2006).

Foto 4. Lichen tundras on 68°N latitude: A – in the Taz Peninsula (neighborhood of Yamburg settlement) – analogue native lichen tundra at Yamal; B – in the Yamal peninsula (Yuribey river besin) – example of present-day lichen tundra under impact of overgrazing.

Foto 5. Criptogamic crusts on polygonal tundras of Yamal: A – вид сверху; B – вид с поверхности земли.

Table 4. Occurrence and abundance of *Cladina* lichens in Yamal tundra in the beginning of 20-th century and at present time

Parameters	Lichen tundras	Dwarf shrub, moss-dwarf shrub (former lichen) tundras	
	Yamal 1933 (Andreev, 1934)	Subarctic tundra zone	
		Southern (2004-2005)	Northern (2005-2006)
Cladina stellaris			
Occurrence, %±SD	≈ 90	0,5±0,7	0,3±0,5
Cover, %	50	0	0
Height, sm	5-7	0,5-1	0,5-1
Mass, ton/hectare	3	0	0
Cladina arbuscula			
Occurrence, %±SD	≈ 90	49,4±11,2	73,7±12,9
Cover, %±SD	20-40	7,5±3,1	3,4±1,9
Height, sm±SD	3-4	1,8±0,4	1,5±0,4
Mass, ton/hectare ±SD	2	0,15±0,04	0,1±0,06
Cladina rangiferina			
Occurrence, %±SD	≈ 80	63±12,4	58±15,2
Cover, %±SD	20-30	6,9±2,4	1,5±0,4
Height, sm±SD	4-5	2,1±0,4	1,8±0,6
Mass, ton/hectare ±SD	2	0,09±0,02	0,17±0,03

Under overgrazing influence the structure of lichen communities has changed. In the beginning of 20th century the lichens of genus *Cladina* (*C. stellaris, C. rangiferina, C. arbuscula*) were basic dominants (edificators) of communities as it is observed now on Taz peninsula in the absence of grazing. On the Yamal at present days *C. stellaris* has almost completely disappeared, the abundance of two others species has considerably decreased (table 4). Instead of them badly or not foraged species of genus *Sphaerophorus, Alectoria, Flavocetraria Thamnolia* are expanded. Wide distributions have received foliose and crustose forms of lichens. Polygonal lichen tundra, small hillocks in lichen hummock tundra and knolls in lichen flat hummock bogs practically everywhere has lifeless soil-grey colour which is created by cryptogrammic crusts in combination with crustose and foliose lichens replaced reindeer moss species (Foto 5).

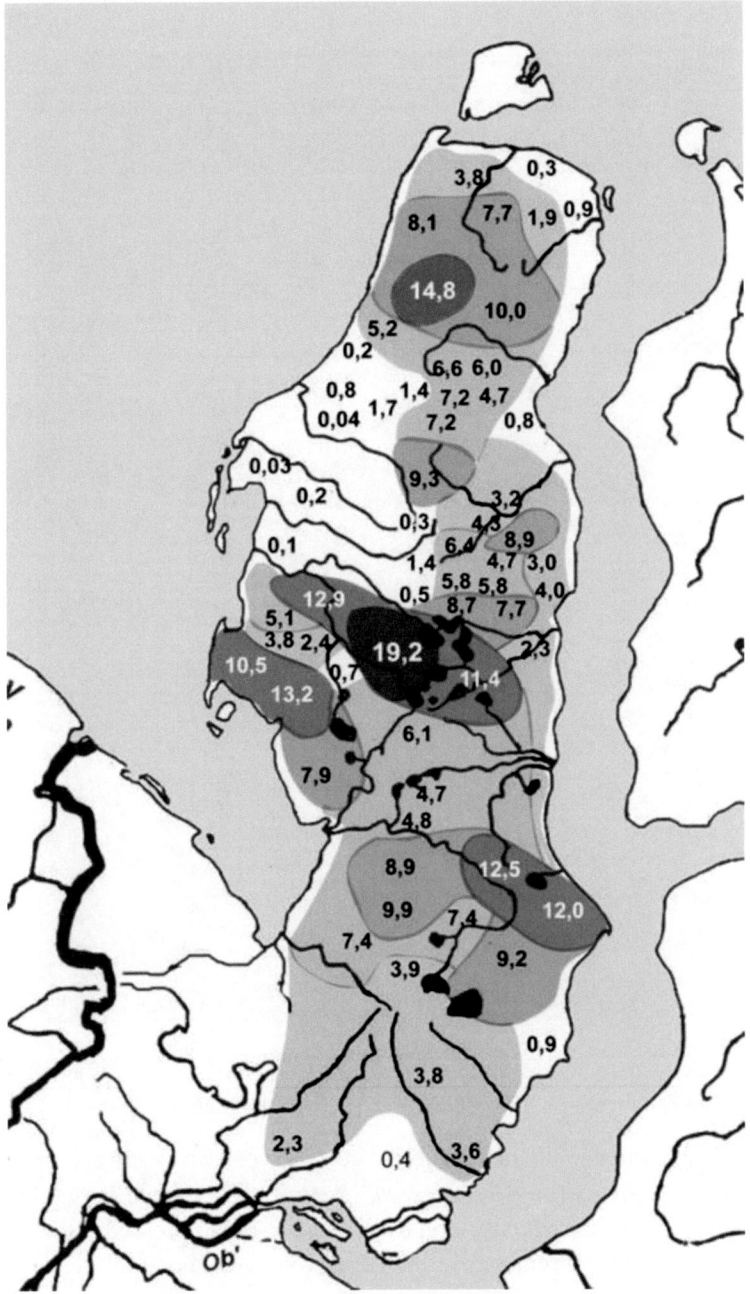

Figure 9. Map-scheme of sand deflation outcrops on Yamal which formed under reindeer grazing impact (arabic numerals is % of sand deflation outcrops from space of land).

Foto 6. Sand deflation outcrops on reindeer pastures of Yamal (Western Yamal, Northern Subarctic Tundra): A – **view from above**; B – view from the ground surface.

Under the influence of trampling the moss cover also is injured and dies off. Crustose lichens of genus *Ochrolechia, Pertusaria, Mycobilimbia* and others settle on died mosses, but grasses and dwarf shrubs are oppressed. Practiced going round of herds on a pasture destroys the vegetative cover completely.

Because of Yamal constituent stratums features (friable and strong sandy) the disturbance of vegetative cover under the influence of reindeer trampling promotes strengthening of deflationary processes. Sandy uncoverings on reaching some critical size become sources of the sand carried on the nearest territories. Gradually the grounds desertification is occurred. Now extensive sand uncoverings exist everywhere on terraces of river-banks and lakesides, on uplands of peninsula (Foto 6). In same places of Middle Yamal its areas reaches to 19% from land area (Figure 9). On the average for peninsula in various tundra subzones the sand uncovering area is similar and now about 5,5% (table 5) that is comparable to the area of lakes on Yamal.

Table 5. Average share of sand deflation outcrops (% from total area of overland ± SD) which forming under grazing impact at different part of Yamal.

Share of sand deflation outcrops	Subzones of tundra			Total for peninsula
	Southern	Northern	Aarctic	
	6,2±3,9	5,6±4,7	4,8±3,9	5,4±4,3

Recovery of grassy plants in particular sedges and cotton-grasses after their damage occurs at one species (for example, *Eriophorum polystachyon, E. russeolum, Carex arctisibirica*) due to growing again the damaged leaves, at others (for example, *C. aquatilis, C. globularis*) – due to an emergence of new shoots from "sleeping" buds which in norm will shoot only the following season. If plants are damaged fully at the beginning of vegetative season the its recovery is completed (> 90%) after 25-30 days at the first group of species, the phytomass of the second group species is restored only on 46% towards the end of season (Peshkova, 1977). At that a depletion of plastic substances reserves in rhizomes has occurred in any case. If plants have not time to fill them in a current vegetative season it affects on their production in the next year. At practiced system of grazing on Yamal when feeding and trampling of vegetation are occured repeatedly the phytomass decreases gradually every year.

Our observations to vegetation on isolated from reindeers plots are demonstrated that in 11 years after experiment beginning the general resources of aboveground phytomass on them has increased in 1,7–1,9 times in comparison with adjoining territory where pasture load remained. However reliable distinctions were observed only on one of plots (table 6). On another the tendency to increase of the general phytomass resources has appeared only ($p = 0,1$). At analyzing of separate structural elements of phytomass it is necessary to note the increasing of dwarf shrubs resources on both isolated plots and lichen masses on one of them. I.e. in 11 years after stopping of the grazing the first signs of tundra vegetation recovery have revealed only.

Table 6. Resources and structure of aboveground phytomass (M±SD г/м²) on ungrazed and grazed plots in dwarf shrub-lichen-moss tundras after 11 years period of grazing pressure absence.

Structural elements of phytomass	Site 1				Site 2			
	without grazing	under grazing	t-test	p	without grazing	under grazing	t-test	p
Dwarf shrubs	22,1±7,8	4,5±3,0	4,65	0,001	13,1±6,9	0±0	4,20	0,002
Sedge and grasses	10,9±3,5	16,3±29,4	0,41	0,69	4,2±7,5	9,6±20,5	0,55	0,60
Lichens	89,6±25,8	51,8±38,2	1,82	0,10	252,5±140,8	90,5±43,5	2,45	0,03
Mosses	103,4±70,9	57,9±38,1	1,25	0,24	94,4±70,7	87,4±86,7	0,13	0,80
Total phytomass	225,9±90,4	130,7±71,8	1,84	0,10	364,2±127,7	187,5±99,2	2,43	0,04

CHANGE OF TERRESTRIAL VERTABRATES FAUNA

The analysis of the literature and our own observation allow to single out many species of vertebrate animals for which in latter 60-70 years was noted changes of number and distribution (table 7). The reasons of this change can be different.

Table 7. List of nesting species for which evident change of number and border of distribution on Yamal peninsula in letter 60-70 years were occurred.

Species	Number	Area of distribution	Species	Number	Area of distribution
Birds:					
Whooper Swan *Cygnus cygnus*	+3	0	Bewick's Swan *C. bewickii*	+3	0
Greylag Goose *Anser anser*	−3	−3	Taiga Been Goose *A. f. fabalis*	−3	0
Lesser White-fronted Goose *Anser erythropus*	−2	0	Mallard *Anas platyrhynchos*	+1	+1

Table 7. (Continued)

Species	Number	Area of distribution	Species	Number	Area of distribution
Garganey *A. querquedula*	+1	+1	Velvet Scotter *Melanitta fusca*	−1	−2
Pallid Harrier *Circus macrourus*	+1	+3	Rough-legged Buzzard *Buteo lagopus*	−1	0
Gyrfalcon *Falco rusticolus*	−2	?	Kestrel *F. tinnunculus*	+1	+1
Hobby *F. subbuteo*	?	+1	Whimbrel *Numenius phaeopus*	+1	+1
Ringed Plover *Charadrius hiaticula*	+2	0	Little Ringed Plower *Ch. dubius*	+1	+1
Great Snipe *Gallinago media*	−3	?	Common Sandpiper *Actitis hypoleucos*	+1	+1
Pomarine Skua *Stercorarius pomarinus*	−2	0	Long-tailed Skua *S. longicaudus*	−1	0
Black-headed Gull *Larus ridibundus*	+1	+2	Little Gull *L. minutus*	+2	+1
Snowy Owl *Nyctea scandiaca*	−2	−1	Rock Dove *Columba livia*	+2	+1
Meadow Pipit *Anthus pratensis*	+2	+2	Yellow Wagtail *Motacilla flava*	?	+1
Fieldfare *Turdus pilaris*	?	+1	Song Thrush *T. philomelos*	?	+1
Redstart *Phoenicurus phoenicurus*	+1	+1	Willow Tit *Parus montanus*	?	+1
Starling *Sturnus vulgaris*	−3	−3	Greenish Warbler *Phylloscopus trochiloides*	+1	+2
Hooded Crow *Corvus cornix*	+3	+1	Rook *C. frugilegus*	−3	−3
House Sparrow *Passer domesticus*	0	+2	Tree Sparrow *P. montanus*	0	+1
Chaffinch *Fringilla coelebs*	?	+1	Brabling *F. montifringilla*	+1	+1
Lapland Bunting *Calcarius lapponicus*	−2	−1	Snow Bunting *Plectrophenax nivalis*	+1	+1
Mammals:					
Lemmings *Lemmus sibiricus Dicrostonyx torquatus*	−3	0	Middendorf's vole *Microtus middendorfii*	+2	+1

(+ increasing, − decreasing; 0 − relative stability; in marks − maximum 3)

Unintelligible reasons of changes. It is possible to add the indicated number of species (Table 7) by a whole series of Siberian birds for which after expansion of their area to the west had advancement to the north too. For example, Arctic warbler (*Phylloscopus borealis*) was fixed in Finland in 1909 (Formozov, 1964), near Lake Ladoga it began to appear regularly after 1971 (Malchevsky, Pukinsky, 1983). At the south of Yamal it was noted at latitude of Salekhard (66°42' N) in the end of 19 century (Derugin, 1898) but to the north it was absent right up to 1941 (Kucheruk et al., 1975). However already in 1978 Arctic Warbler was found near Sokhonto Lake at latitude 69°06' N (Danilov et al., 1984). Bluetail (*Tarsiger cyanurus*) which has approached Kola Peninsula in 1937 (Malchevsky, Pukinsky, 1983) and Finland in 1949 (Formozov, 1964), at the south of Yamal has been observed at the end of 1970th (Danilov et al., 1984). Olive-backed Pipit (*Anthus hodgsoni*) hardly approached to latitude 64° N in 1904 (Shukhov, 1915) but after 2002 it has regularly founded in Low Ob'River area at latitude 66°42' N (Golovatin, 2002). Appearance in the fauna of Yamal of such birds as Pintail snipe (*Gallinago stenura*), Petchora pipit (*Anthus gustavi*), Common rosefinch (*Carpodacus erythrinus*), Pallas's Reed Bunting (*Emberiza pallasi*) is connected probably also with movement of borders of their area to the north-west in some time. But about concrete terms of their penetration to Yamal are difficultly to discuss for lack of ornithological data. Anyway it is difficult to connect appearance of listed Siberian species with climate warming or anthropogenic changes. It is particular displays for a long time ago beginning process of advancement of the east-asian elements of fauna to Europe.

At some species changes of number or distribution on Yamal are reflection of general tendencies occurring simultaneously in many parts of their areas. For example occurring everywhere number decreasing of Lesser White-fronted Goose (Morozov, 1995; Morozov, 1999), Velvet Scotter (Mineev, 2003), Great Snipe (Malchevsky, Pukinsky, 1983; Bakka, 1990; Askeev, Askeev, 1999), Starling (Moller, 1992; Robinson et al., 2005; Sotnikov, 2006) has affected also on the populations of these species on Yamal peninsula. Immediate reasons of such changes are not looked through. At least it is not pronounced connection either with climate warming or with anthropogenic transformation of landscape. Probable explanation may be a specificity of species reactions on obscure while factors. In particular M.A. Menzbir (1882) arguing on fate of Great Snipe prior one century to the designated reduction of its number specified that unlike of Snipe (*G. gallinago*) it should disappear gradually in connection with ploughing up of meadows, drainage of tuussock bogs, easy and unlimited hunting.

Cyclical fluctuations of numbers. In certain cases changes are a part of long term cyclic fluctuations of bird numbers. Usually it is difficultly to establish Cyclical fluctuations may be difficalt ascertained in connection with relatively short period of observations. But when fluctuations are relatively short-term the connection of numbers movement with its natural cycles becomes obvious that is accompanied by corresponding variations of species area borders. For example for Red-breasted Goose (*Rufibrenta ruficollis*) it is peculiar cycles with periodicity in 12-14 years and 3-4 years depending on weather conditions of reproduction season (Krivenko, 1991). In the second half of 1990th the regular increase of total numbers of this species had been beginning (Syroechkovsky, 1995). It had led to restoration of borders of breeding area on periphery: at the south of Yamal in upper reaches of Khadyta-yaha river (67°19' N) in 1997 (Mechnikova et al., 2005), at the south-west on Erkuta-yakha and Payuta-yaha rivers (68°11' N) in 2001 (Sokolov et al., 2001; Sokolov, Sokolov, 2005), in the north-west on the Naduy-yaha river (70°37' N) in 2006 (Shtro, Sokolov, 2006). In traditional places of breeding on the Yuribey river to 2004 both total numbers of adult birds and numbers of broods had increased (Golovatin et al., 2004). However already in 2005 repeated inspection of the river showed that numbers of Red-breasted Goose had reduced in 3 times here. Birds nested on places of former colonies but only by separate pairs.

For other example numbers of swans wintering in the north of Europe has expressed cyclic dynamics with periodic liftings: **for Bewick's Swan everyone 6-9**, for Whooper Swan – 3-5 years (Mineev, 2003). Most likely besides more long cycles are in existence. So during the period from the end of 1930th till the end of 1950th **Bewick's Swans were enough numerous** on Yamal (Pugachuk, 1965; Kucheruk et al., 1975). Its frequency in coastal part of peninsula was 22,5±0,8 individuals/10 km of route. Then population numbers of birds wintering in the Western Europe decreased greatly but from the middle of 1980th the increase has begun again (Beekman, 1997). In this time on Yamal the numbers of this swans has increased in ten and more times. So when Yuribey river (the central part of area of species on Yamal peninsula) were investigating in 1982-1986 the **frequency of Bewick's Swan in river valley** was estimated in 0,18±0,04 individuals/10 km. From time to time single birds, detached pairs and more rarely small groups of 3-4 individuals were met. In 2004-2005 the frequency was 4,2±0,4 individuals/10 km of river valley and nearly to sea – 10,0±1,3 individuals.

One more example is Greylag Goose. According to remains of bones in ancient settlements it has been distributed in forest-tundra of Western Siberia in 17 – first third 18 century (Kosintsev, 2006). In the end of 19 century it was found as breeding along Ob river to most lower part – mouth of Schyuchya river (66°48'

N) (Finsch, 1879; Derugin, 1898). In 1930-1940th the steady tendency of essential decrease of total numbers of species was appeared. It clashed with phase of low water in the central part of area of the species – in steppe zone. The species has disappeared on Southern Yamal. In Ob flood-lands the border of its distribution has receded to 66°10' N and after middle of 1960th it has ceased to nest here too (Danilov, 1969). From the beginning of 1980th a new phase of the hydroclimatic cycle began and water level started to increase. After this the increase of the total goose numbers had began especially in forest-steppe and in the north of steppe zone which had been disturbed by meliorative transformations to a lesser degree (Yerokhov, 2003). This tendency go on at present time too. As a result Greylag Gooses visitations on the north become more frequent, along Ob River up to 68°28' N (Ryabitsev et al., 1995). However at present time the nesting of the species on Ob River probably is restrained by activity and disturbance of the increased people population.

The faunistic reorganizations connected with climate change. A movement to the north of the species nesting southward is usually connected with climate warming. During last 75 years in the north of Western Siberia four periods of appearance of «southern» species have been (Figure 10). These periods were conditionally dated for years when early springs were frequent enough. The conditional character is explained by that usually to record first appearance of new species is not succeeded at once, but most often with some delay. Especially it concerns three species of the first period. They had been found after 24 years interruption in investigations (the previous visiting of Yamal was realized by I.N. Shuhov in 1913). Quite probably those birds had appeared in warm springs of 1920-1924 but there was nobody there to find them. Penetration to the north of group of species in 1978-1982 can be related to the short period of early springs in 1976-1982.

N.N. Danilov (1966) analyzing the nature of the birds distribution in Subarctic came to conclusion that penetration of species into high latitudes is limited by features of gonads functioning and conditions at habitats during an arrival and beginning of the nesting. «Biological clocks» of every species are adjusted so that the moment of its gonads maturing is concurred with the convenient period for reproduction. For typical subarctic species this time is shifted for later calendar terms in comparison with the species penetrating into Subarctic region. In years with usual northern spring "southerners" breeding relatively early clash with absence of necessary habitat conditions at their moment of physiological readiness for reproduction and that restrains the bird distribution to the north. In early springs a shift of the phenological phenomena occur and that gives chance for birds to nest in high latitudes.

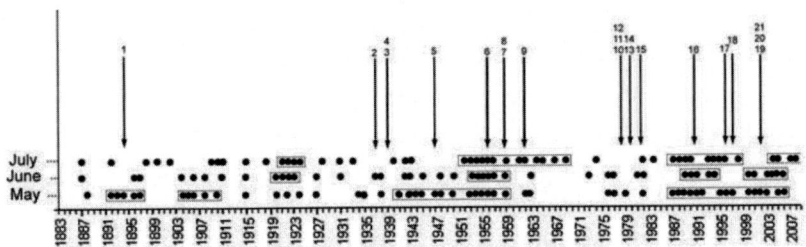

Figure 10. Established dates of first nesting of «southern» birds at the south of Yamal on the background of years with warm spring and summer. Contours mark periods of frequent recurrence of deviation of the temperate from many years average. Numerals indicate: 1 – House Sparrow; 2 – Brabling. 3 – Whimbrel, 4 – Meadow Pipit, 5 – Arctic Warbler (Kucheruk et al., 1975); 6 – Mallard, 7 – Common Sandpiper (Dobrinskii, 1959); 8 – Kestrel (Danilov, 1969); 9 – Garganey (Danilov, 1965); 10 – Bluetail, 11 – Redstart, 12 – Greenish Warbler, 13 – Little Gull, 14 – Song Thrush (Danilov et al., 1984); 15 –Hobby (Kalyakin, 1998); 16 – Chaffinch (Golovatin, 1995); 17 – Little Ringed Plower (Paskhalny, Sinitsyn, 1997); 18 – Pallid Harrier (Morozov, 1998); 19 – Olive-backed Pipit, 20 – Willow Tit (Golovatin, 2002); 21 – Rock Dove (Paskhalny, 2004).

However it is interesting that almost all new species after their penetrating to the north have remained here for many long years and have nested independently from following seasons will be warm or cold. The avifauna of Southern Yamal has a lot of species for which the basic area is far to the south: Slavonian Grebe (*Podiceps auritus*), Shoveler (*Anas clypeata*), Tufted Duck (*Aythya fuligula*), Goldeneye (*Bucephala clangula*), Smew (*Mergus albellus*), Spotted Crake (*Porzana porzana*), Great Snipe (*Gallinago media*), Terek Sandpiper (*Xenus cinereus*), Black-headed Gull (*Larus ridibundus*), Common Tern (*Sterna hirundo*), Skylark (*Alauda arvensis*), Siberian Accentor (Prunella *montanella*) and others. Probably these birds have penetrated here into the previous warm periods which presences are accurately seen in dendroclimatic reconstruction of climate (Figure 4). A confirmation may be by the picture of emergences of some listed species in regions where ornithological investigations are being carried out for a long time, for example in Baltic. Spotted Crake, Black-headed Gull, Shoveler, Tufted Duck began to assimilate new habitats here only in 1880th (Isakov, 1952; Kumari, 1957) after the so-called «small glacial period» during 1550 – 1700-1800 years (dates by Lamb, 1977). It is quite probable that at this time they have penetrated to Southern Yamal too.

In most cases the process of assimilating of new territories is enough typical: appearance of the first nesting birds (regular visitations of «scouts» may precede that but may be absent), then increase of numbers and widening of nesting area. For example the first mentions about Whimbrel (*Numenius phaeopus*) in the north of Western Siberia fall on 1939 when there were only 4 meetings in the basin of Schyuchya river, after 34 years (since 1973) it already quite usual species (Kucheruk et al., 1975) with enough stable density of nesting.

At the same time after first «wave» of numbers increasing and widening of the area a stagnation may occur and even reduction of population numbers than after that a new wave may begin. The example of Pallid Harrier is demonstrative. The beginning of the first wave of species advancement to the north was noted on a boundary of 1880-1890th (Formozov, 1959) when it became usual in the Central Russia (the south of Tula province) at latitude near 53°20' N. In 1913 Pallid Harrier was found as nesting under Saint Petersburg – 59°41' N, in 1935 has advanced to 60°40' N (Malchevsky, Pukinsky, 1983). At the east of the European Russia for these years the northern border of its area has reached approximately latitude 58°40' N (Plessky, 1948). In the 1950th reduction of numbers of the species has begun and in 1970th it became evident in the central part of the area, in particular in the south of Bashkiria (Ilyichev, Fomin, 1979). In 1990th years the new wave of Pallid Harrier advancement on the north has begun in the north-east direction, as a result the species has penetrated into Yamal. In 1991 it was founded in Bolshezemelskay tundra (68°38' N) (Morozov, 1997), in 1998 – in the south of Yamal (67°29' N) (Morozovs, 1998), in 2001 and 2002 – at the Polar Urals (Golovatin, Paskhalny, 2005) and on the south-west of Yamal (68°12' N) (Sokolov et al., 2002), in 2004 – in middle part of the peninsula (68°46' N) (Golovatin et al., 2004).

Disappearance of «new» species occurs only in single cases later the some time after penetration and over reasons not connected with weather. As reliable examples it is possible to mention Starling and Rook. The history of the last species is the most evident as occurred during rather short time interval. The first nesting of Rooks (12 pairs) was fixed in 1975 near Labytnangi on transmission towers. But in 1982 a colony had 8 pairs only, in 1983 – 7, in 1984 – 1 (Paskhalny, 2004). The next years, i.e. 10 years after the first registration, nesting birds did not meet.

The faunistic changes connected with anthropogenic influence. In isolated instances the reason of disappearance or reduction of species numbers is the direct persecution. Before beginning 1960th Capercaillie (*Tetrao urogallus*) was regularly met in a forest-tundra zone of Yamal. It was hunted even in distant forest island in the valley of Khadyta-yaha river (Danilov et al., 1984). During coldness in 1960th – first half 1970th Capercaillie had disappeared here.

The basic nesting place of Gyrfalcon in the region is valley of Schyuchya River. After building of Obskaja-Bovanenkovo railway and concomitant road these places became accessible to the poachers practising capture of this rare falcon for contraband sale. As a result the number of nesting Gyrfalcon was distinctly reduced from an order 10-15 to 5 pairs (Mechnikova, Kudryavtsev, 2005).

At the same time growth of the population and formation of a system of settlements promoted area widening of a whole series of species, first of all synanthropic such as sparrows, Rock Dove and semisynanthropic Snow Bunting. At the end of 19 century House Sparrow had reached Berezovo town (63°54' N) and Tree Sparrow just north – to Kushevat village (65° N), but for the winter they flew away (Finsh, Brem, 1882). On Yamal at this time Obdorsk (today Salekhard – 66°31' N) was unique village with wooden building, other settlements represented the compact camps of indigenes consisting from several chums. The sparrows began settled in Obdorsk after 20 years (Deryugin, 1898) in the period of early springs (Figure 10). By the early 1980th with the advent of settlements along Ob River they began to nest regularly on the north to settlement Yarsale (66°51' N). In the 1990th periodically nascent centers of breeding have appeared in working settlements on gas fields in the Southern, Northern and even in the Arctic tundra. Now aggregations of House Sparrow are found in all constantly existing large settlements including the most northern – Bovanenkovo (70°21' N) and Kharasavey (71°05' N). There are data about encounters of the bird in small settlements (trading stations, bore hole) up to northern part of the Arctic tundra subzone and about wintering of visited birds in small fishing settlement Yaptik-Sale (69°23' N) (Paskhalny, 2004). Along with active moving of birds they was bringed by vehicles: motor ships, air-freighters and helicopters.

All stable settlements of Rock Dove in the north of YNAR also were formed on the basis of the bringed birds and closely connected with appearance of high-rise buildings where on warm garrets the birds have found suitable places for overnight rest and nesting. The first aggregation in the north of Western Siberia was found in 1984 in Nadym (65°31' N). At the south of Yamal in Labytnangi (66°39' N) in 2002 about 10 birds settle down after construction of high-rise brick building with the warm ventilating top storey having on perimeter reach-through holes (Paskhalny, 2004). Now (6 years later) after increase of high-rise buildings number the bird population has grown approximately to 250 individuals.

Snow Bunting in natural environment lives in mountains of Polar Urals up to southern border of YNAR (Golovatin, Paskhalny, 2005) and on Yamal peninsula only in northern half of Arctic tundra subzone (above 72° N) mainly at coasts (Paskhalny, 1985). Its moving on peninsula territory was from the north to the south and from mountains in plain as increase of settlements number and creation of new habitats in the form of the quarries, abandoned constructions and railway embankment. In 1961 it has appeared at settlement Mys Kamenniy (68°28' N) (Leonovich, Uspensky, 1965), in 1987 – in settlement New Port (67°41' N), in 1989 – in working settlements on Bovanenkovo gas field (70°21' N), in 1998 – on objects of building railway Obskaya-Bovanenkovo (67°04' N) (Paskhalny et al., 1998).

For nonsynanthropic «southern» species on Yamal the known rule of increase of synanthropization as approaching to borders of area with growth of severity abiotic factors (Klausnitzer, 1987) are operated. For example the most northern point of Yellow Wagtail nesting (compact group of three pairs) in the middle part of Yuribey river (68°38' N) related to anthropogenic site – abandoned derrick. Wood Sandpiper (*Tringa glareola*) on the limit of distribution (70°21 ' N) was found in territory Bovanenkovo gas field. The average nesting density of Meadow Pipit was reliable higher in the field territory than beyond the bounds of one (1,33 pair/km^2±1,13 SD against 0,27±0,29; t = 2,24, p = 0,04). At northern border of Tundra Been Goose distribution on territory of the Kharasavey gas field (71°11' N) the main part of broods (8 of 9) has been concentrated to settlement outskirts (Foto 7).

Foto 7. Broods of Been Goose on the outskirts Kharasavey settlements (Arctic tundra – 71°11' N).

Foto 8. Nest of Fieldfare under awning of building at the town Labitnangi (66°46' N) – the example of unusual nests organization on the border of species area.

Foto 9. Recently fledged young of Hooded Crow at the Kharasavey settlement (Arctic tundra).

Some species in the north of area build their nests in unusual places. For example Fieldfare nests in natural habitats on trees, in mountains – on ledges and in niches of rocks (Golovatin, Paskhalny, 2005). On northern border of the area we founded its nests in the old projector on pillar, on a ledge under the awning of house-top (Foto 8), on the ledge of steel beam under the flooring (Paskhalny, 2004). In tundra outside of settlements the overwhelming majority of nests was built on rocky and macadam quarries, on metal and wooden bridge footings. The most northern nest on Yamal (68°38 ' N) has been located on the abandoned derrick (Golovatin et al., 2004).

Other example. Till recently the most northern nest of Hooded Crow was founded in 1974 in the northern part of Southern subarctic tundra subzone in vicinities of settlement Mys Kamenniy (68°40' N). It was in the flood-lands of Nyurma-yaha river and located on willow top in 2 m from the ground (Danilov et al., 1984). Spring visitations of particular birds and flocks occured further away to the north to the trading station Tambey (71°31' N) in 1974 and Marre-Sale (69°43' N) in 1976. In the 1980th the crow in the north of its area began to assimilate new places for nesting – to build its nest on wooden and metal constructions (Paskhalny, 2004). In 2008 there was fixed case of the crow nesting in the Arctic tundra in outskirts of settlement Kharasavey (71°11' N), where birds built its nest on the metal construction of oil products storehouse (Foto 9) (Golovatin, Sokolov, 2008).

Changes of the animal population connected with reindeer grazing. As we already noted above overgrazing of reindeers on Yamal has included all territory of peninsula and became the reason of large-scale changes of vegetative cover and strengthening of deflationary processes. The transformation of habitats caused by degradation of vegetation owing to unlimited pasturage of cattle always led to essential changes of animal numbers and distribution on open landscapes (Formozov, 1959, 1962; Sandford, 1976). Yamal did not become an exception.

Here before 1990 considerable explosions of lemmings abundance (synchronously *Lemmus sibiricus* and *Dicrostonyx torquatus*) occurred with periodicity of 3-4 years that is typical for the Arctic areas. For last 18 years the numbers of lemmings has never reached former high values, only twice to an average level though the cyclicism of raisings has remained (Figure 11). At that raisings began to have local character and do not cover as before all or the most part of peninsula territory. Connection of these disturbances with overgrazing is obvious since they has concur with high growth of reindeers livestock and with beginning of uncontrolled grazing. Against decrease in an abundance of lemmings number of other tundra rodent – **Middendorf's vole** (*Microtus middendorfii*) who in years of peak is intensivly settled on all tundra has several times increased occupying habitats where it did not meet before.

In the natural situation repeated fluctuations of lemmings numbers affect on all links of tundra biocenosis and, first of all, on predators. Fertility and success of their reproduction sharply increase at a high abundance of these rodents. Falling of lemming numbers forces to be switched predators to secondary forages (voles, birds). But transition to replacing kinds does not fill that reproductive effect which exists at a food of predators by lemmings, even at an abundance of voles. Especially it is peculiar to predators, specialised on a food by lemmings such as Arctic fox, Snowy Owl, Pomarine Skua and has been noticed at them earlier (Danilov et al., 1984; Shtro, 2003; Golovatin, Paskhalny, 2005a; Morozov, 2005). Therefore it is no wonder that for last 15 years did not arrive any authentic data on mass nesting Snowy Owl and Pomarine Skua on Yamal. The density of nesting of Snowy Owl to 1991 reached 0,08-0,11 pairs/km^2 (Golovatin, Paskhalny, 2005a), after nests did not find. Dense settlements of Pomarine Skua were marked locally and only in the Arctic tundra. In the southeast of the subzone they nested in lower reaches of the river Venujeuo (71°07' N 72°27' E) in 1991 with density 2,36, in 1993 – 0,12 and in 1994 – 0,68 pairs/km^2 (Ryabitsev, Taylor, 2001). In 1997 small aggregation with density of 1,8 pairs/km^2 it was revealed us in seaside tundra near Kharasavey (71°10' N 67°50' E). Slight increase of numbers of Siberian lemming to average values has been provided by successful under snow reproduction but in the summer they did not breed. In 90 km to the south – around settlement Bovanenkovo (70°21' N 68°21' E) they were not absolutely. In 2004 nesting of Pomarine Skuas on island Beliy (73°16' N 70°42' E) (Dmitriev, Emelchenko, 2005) has been fixed. Here the reindeer grazing are not present and also peaks of lemming numbers to considerable sizes still happen. This serves as acknowledgment of overgrazing influence on dynamics of their abundance.

Predators, less specialized on a food by lemmings – Rough-legged Buzzard and Long-tailed Skua nest in absence of lemmings only at a high abundance of voles. But thus their density **much more low ($p \leq 0,001$) than in «lemmings» years** (table 8).

Grazing influence on the other groups of birds we illustrate on an example of basin of Juribey river on Average Yamal where materials of censuses have the greatest volume (in different years of 6-13 plots in the general size 120-500 km^2) and are long in time (1982-1986, 1991, 1997, 2004-2005) that allows to spend representative comparison for different years (Table 9).

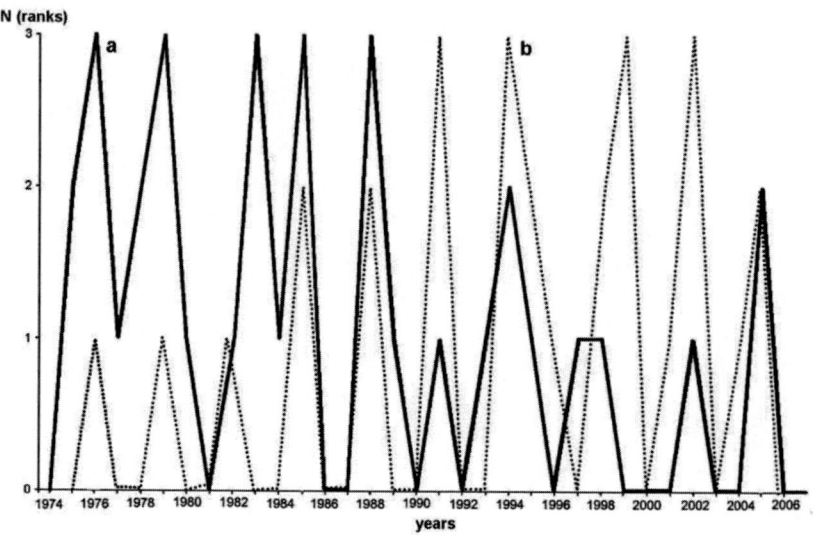

Figure 11. Number dynamics of lemmings (a) and Middendorf's vole (b) on Yamal (in marks, lemmings – by Shtro, 2003; Shtro, Sosin, 2004; authors data; Middendorf's vole – by Balakhonov et al., 1997; authors data).

Table 8. Average nesting density (pairs/10 km² ± SD) of Rough-legged Buzzard and Long-tailed Skua at years of abundance of lemmings and voles on Yamal.

	Abundance of lemmings	Abundance of voles	t-test, (f)
Rough-legged Buzzard	5,7±2,3	2,4±1,2	4,61 (24)
Long-tailed Skua	2,2±1,2	0,7±0,5	4,41 (21)

After 1990 number at many groups of terrestrial birds especially on the watershed was reduced. Decrease in an abundance of such typically tundra birds as geese, the most mass species of water-fowl – Long-tailed duck, waders and Lapland bunting is indicative. At geese it was expressed in reduction of number of broods and molting birds on the river. Number of Long-tailed duck has decreased on a watershed in 2 times, hygrophilous waders – almost in 3 times, not only on a watershed, but also in a flood-plain. The density of Lapland Bunting on watersheds was reduced more than in 5 times, in flood-plain almost in 2,5 times. It was reflected in a state of the species near southern border of area. So in vicinities of Labytnangi (66°46' N) in 1970th its density of nesting made 2,6-13,6 pairs/km^2 of tundra (Danilov et al., 1984), now – it here has practically disappeared – less than 0,1 pairs/km^2 (our data). On the other hand owing to landscape desertification the area of habitats of the characteristic inhabitant of mineral arenas – Ringed Plover has extended. It has led to increase in its numbers more than in 2 times. Growth of numbers of Bewick's Swans, most likely, does not concern to grazing, and is reflexion of the steady process occurring with this species within last decades (Beekman, 1997).

CONCLUSION

As show results of dendrochronologycal researches on Yamal and Polar Urals during warming VIII-XIII centuries there was an intensive expansion of tree vegetation which was replaced by its deviation at a cold snap in the end of XIII – the beginning of XX century. Arheobotanical analysis confirms that in medieval warming the border of the Subarctic tundra subzone on Yamal has been considerably shifted to the north. Modern warming is comparable with one XII-XIII centuries but the tree vegetation yet has not reached the climatic caused limit of XIII century. The evidence of any scale changes of a vegetative cover of the tundra connected with climatic cycles of smaller periodicity or last warming is absent.

Table 9. Average density (individuals/km^2 ± SD) of birds at basin of Yuribey river (Middle Yamal) before and after 1990 (significant difference: * p ≤ 0,05, ** p ≤ 0,01, *** p ≤ 0,001, f = 7).

Species and groups of birds	watershed			flood-lands		
	before 1990	after 1990	t	before 1990	after 1990	t
Arctic Loon Gavia arctica	0,26±0,16	0,32±0,11	0,60	0,90±0,62	0,75±0,16	0,59
Bewick's Swan Cygnus bewickii	0,01±0,01	0,08±0,03	3,54**	0,02±0,02	4,58±0,57	15,86***
All geese	0,39±0,14	0,31±0,28	0,58	1,68±0,61	0,95±0,30	2,40*
Pintail Anas acuta	0,18±0,16	0,04±0,02	1,55	3,00±1,42	2,70±0,86	0,40

Table 9. (Continued)

Long-tailed duck Clangula hyemalis	1,67±0,26	0,83±0,56	3,00*	6,40±3,27	5,67±0,35	0,50
Scaup Aythya marila	0,33±0,29	0,31±0,19	0,13	0,76±0,43	1,17±0,40	1,57
Willow Grouse Lagopus lagopus	28,38±15,32	10,63±1,24	2,58*	28,64±21,92	35,72±11,73	0,64
Plovers Pluvialis sp.	1,10±0,45	1,10±0,12	0,01	0,37±0,40	0,18±0,09	1,05
Ringed Plover Charadrius hiaticula	0,51±0,32	1,22±0,03	5,02**	0,24±0,54	0,25±0,19	0,05
Other waders	30,00±8,29	11,38±0,26	5,02**	85,36±19,15	28,52±4,28	6,48***
Lapland bunting Calcarius lapponicus	40,06±24,80	7,45±3,50	3,68**	78,75±25,52	33,06±18,06	4,13**
Other passerine	88,94±54,48	45,71±13,68	2,18	82,42±34,57	121,08±22,3	2,66*

The present technogenic changes of a vegetative cover connected with prospecting and exploitation of hydrocarbonic fields on peninsula have local character and are insignificant on the area. Penetration on the north of some southern plants in process of formation of a technogenic landscape is marked.

Now the most powerful and large-scale factor of influence on tundra biocenoses of Yamal became intensive grazing of domestic reindeer which livestock after 1990 has passed in a phase of rapid growth and makes now about 300 thousand heard (at average density about 2,3 ind/km^2). Under grazing effect the native structure of the vegetative cover at peninsula has changed. Lichen tundra has most strongly suffered. In comparison with 1930th its total area was reduced in 3,5-4 times, the prevailing height of a lichen cover has decreased with 3-4 to 0,5-1,5 sm. The composition of dominant species was replaced – instead of fruticose lichens of genus *Cladina* were extended the poorly or not foraging species of foliose and crustose forms. Polygonal lichen tundras on watersheds practically are everywhere presented by cryptogamic crusts in a combination with crustose and foliose lichens.

Resources of grasses in tundra and on bogs have decreased in 1,5-2 times. On places with high and very high pasturable loading the grassy vegetation is foraged against to height 3-5 sm. I.e. "grinning" of tundra, noted in other areas of Arctic regions, on Yamal is not observed. Under grazing influence shrub stands degrade, there is a decrease of aboveground phytomass of shrubs (undersized – almost in 8 times, tall – in 2 times). The share of dwarf shrubs has increased in structure of a vegetative cover, having for reindeer minor food value.

Because of the features of composing layers of Yamal peninsula the disturbance of a vegetative cover under the influence of trampling by reindeer promoted strengthening of deflationary processes. On Middle Yamal sandy exposures reache locally of 19%, on the average on peninsula – about 5,5% of the area of a land.

After change of vegetative cover structure there was a disturbance of natural dynamics of an abundance of lemmings. Characteristic periodic outbreaks of their number after 1990 have ceased to reach former high values. Despite remained recurrence, they began to carry local nidal character and do not cover as before all or the most part of territory of peninsula. Number of other tundra rodent – **Middendorf's vole** thus has several times increased.

Decrease in number of lemmings was reflected in predators. Nesting of birds, specialised on a food by lemmings (Snowy Owl, Pomarine Skua) became isolated. The density of less specialised species (Rough-legged Buzzard, Long-tailed Skua) has decreased and does not reach former values even in days of a high abundance of voles. It was reduced number of other typically tundra birds: geese, Long-tailed duck, hygrophilous waders, Lapland Bunting too. At the same time owing to landscape desertification the abundance of the inhabitant of mineral arenas – Ringed Plover has increased.

In parallel with these large-scale changes against cyclic fluctuations of a climate there is a constant wavy replenishment of peninsula fauna. Penetration on the north of new "southern" species is dated for the periods when frequent enough were early spring. New species, once having got on the north, remain here for many long years and nest irrespective of the subsequent fluctuations of a climate. At a number of species expansion of area borders is connected with the technogenic changes of a landscape caused by development of gas and oil industry in region. As a result there is a gradual increase in total number of species in the fauna list. The number of the disappeared species, in comparison with newcomers, is insignificant. This process of the directed faunistic reorganization, most likely, is reflexion of large long-term climatic cycles – in particular approach of the next warm epoch.

The work was made with the support Project NEC 02.740.11.0279 and RFBR 08-04-01028.

REFERENCES

Andreev, V. N. (1934). The Forage basis of the Yamal reindeer-breeding. *Sovetskoe olenevodstvo*, *1*, 99-164. (In Russian).

Andreyashkina, N. I. (2008). Formation of vegetative cover in the artificial created habitats (Yamal peninsula) In: S. P. Paskhalny (Editor), The Scientific bulletin of the Yamal-Nenets autonomous region. Part 1: Vegetation and fauna of Ural Mountains and Western Siberia. *Salekhard*, *1(53)*, 3-10. (In Russian).

Askeev, I. V. & Askeev, O. V. (1999). Ornitofauna of Republic Tatarstan (the current state abstract). *Kazan'*, 1-124. (In Russian).

Atlas of Yamal-Nenets autonomous region. (2004). *Omsk*, *303*, (In Russian).

Bakka, S. V. (1990). Rare birds of Gorky region. In: Rare species of birds of the center of Nechernozemya region: *Proceedings of the Conference on The current state of rare nesting bird populations of the Nechernozem center of the USSR*, 27-28, November, 1989, Moscow. 30-33. (In Russian).

Balakhonov, V. S., Danilov, A. N., Lobanova, N. A. & Chibirjak, M. V. (1997). The study of number dynamics of small mammals in the south of Yamal. *In: Proceeding of the Conference on the history and current state of fauna on the north of Western Siberia*. Cheliabinsk, Rifey, 43-59. (In Russian).

Beekman, J. (1997). International censuses of the north-west European Bewick's Swan population, January 1990 and 1995. *Swan Specialist Group Newsletter*, *6*, 7-9.

Briffa, K. R. & Jones, P. D. (1993). Global surface air temperature variations during the twentieth century: Part 2. Implications for large-scale high-frequency palaeoclimatic studies. *The Holocene*, *3*, 77-88.

Danilov, N. N. (1965). Birds of the Lower Ob'River and the change in their distribution for last decades. *In: Proceedings of Institute Biology of UD RAS: Ecology of vertebrate animals of the Far North*. Sverdlovsk, *38*, 103-109. (In Russian).

Danilov, N. N. (1966). Ways of the adaptation of terrestrial vertebrate animals to living conditions in Subarctic. Vol. 2. Birds. *In: Proceedings of Institute Biology of UD RAS*. Sverdlovsk, *56*, 1-147. (In Russian).

Danilov, N. N. (1969). Birds of the Middle and Northern Urals. Part 1. History of research of birds at Urals: Orders of Gaviiformes, Podicipediformes, Pelecaniformes, Ciconiiformes, Ansereformes and Falconiformes. *Proceedings of the Ural Division of Moscow Society of Environmental Researcher*. Sverdlovsk, *3*, 3-123. (In Russian).

Danilov, N. N., Ryzhanovsky, V. N. & Ryabitsev, V. K. (1984). *Birds of Yamal*, Moscow. 1-334. (In Russian).

Derugin, K. M. (1898). Travel in a valley of middle and down stream of the Ob River and fauna of this region. In: Proceedings of Sait Peterburg society of naturalists. *Dep. of zoology and physiology*, Vol, 29, 2, 47-140. (In Russian).

Dmitriev, A. E. & Emelchenko, N. N. (2005). 29. Bely Island, the Kara Sea, Russia (73°16′ N, 70°42′ E). Arctic Birds: *Newsletter of International Breeding Conditions Survey*, 7, 12.

Dobrinsky, L. N. (1959). Data about northern limit of distribution of some birds species in territory of Yamal-Nenets national district. *In: Proceedings Salekhard station on fauna of the Priobsky North and its use. Tyumen, 1*, 167-384. (In Russian).

Fedorova, N. V., Kosintsev, P. A. & Fitshju, V. V. (1998). Go away to the hills. *Ekaterinburg*, 1-132. (In Russian).

Finsh, O. & Brem, A. (1882). *Travel to Western Siberia*, **Moscow**. 1-637. (In Russian).

Finsch, O. (1879). Reise nach West-Sibirien im Jahre 1876. Verhandlungen der zoologisch-botanischen Gesellschaft in Wien, Bd XXIX, *Wien., Jahrgang*, II. 128-280.

Formozov, A. N. (1959). About movement and borders distribution fluctuation of mammals and birds. *In: Geography of the population of land animals and methods of its studying*. Moscow, Academy of Science of the USSR Press, 172-194. (In Russian).

Formozov, A. N. (1962). Change of environment of the steppe South of European part of the USSR for last hundred years and some lines of modern fauna of steppes. *In: Researches of geography of animal and vegetative natural resources*. **Moscow, Academy of Science of the USSR Press**, 114-161. (In Russian).

Formozov, A. N. (1964). Plain of Western Siberia and the features of fauna connected with it. *In: Development and transformation of the geographical environment*. **Moscow, Science Press**, 201-221. (In Russian).

Golovatin, M. G. (1995). New data on birds distribution **of Lower Ob'River** region. In: V. K. Ryabitsev (Editor), Materials to distribution of birds in Urals, TransUrals and Western Siberia. *Ekaterinburg*, UD of RAS Press. 12-13. (In Russian).

Golovatin, M. G. (2002). **Interesting ornithological meetings in Ob'River** region. In: V. K. Ryabitsev (Editor), Materials to distribution of birds at Urals,

TransUrals and Western Siberia. *Ekaterinburg*, UD of RAS Press, 92-93. (In Russian).
Golovatin, M. G. & Paskhalny, S. P. (2005). *Birds of the Polar Urals*. Ekaterinburg, Ural State Univ, Press, 1-560. (In Russian).
Golovatin, M. G. & Paskhalny, S. P. (2005). Distribution, numbers and status of Owls in the northern part of Western Siberia. In: S. V. Volkov, V. V. Morozov, & A. V. Sharikov (Eds.), *Owls of the northern Eurasia*, Moscow. 321-331. (In Russian).
Golovatin, M. G., Paskhalny, S. P. & Sokolov, V. A. (2004). Data about birds fauna of the river Juribey (Yamal). In: V. K. Ryabitsev (Editor), Materials to distribution of birds at Urals, TransUrals and Western Siberia. *Ekaterinburg*, UD of RAS Press, 80-85. (In Russian).
Golovatin, M. G. & Sokolov, V. A. (2008). About Hooded Crow distribution in tundra zone of Yamal. In: V. K. Ryabitsev, & V. V. Tarasov (Eds.), Materials to distribution of birds at Urals, TransUrals and Western Siberia. *Ekaterinburg*, UD of RAS Press, *31*, (In Russian).
Gudina, A. N. (1999). Methods of the census of nesting birds: mapping of territories. *Zaporozhie*, 1-241. (In Russian).
Ilyichev, V. D. & Fomin, V. E. (1979). Ornitofauna of Bashkiria and its change in the XX-th century. *Ornithology, Vol. 14*, 83-96. (In Russian).
Isakov Yu, A. (1952). Subfamily Anatinae. In: G. P. Dementyev, & N. A. Gladkov (Eds.), *Birds of Soviet Union*. **Vol. 4. Moscow: Soviet science Press.** 344-636. (In Russian).
Kalyakin, V. N. (1998). Birds of Southern Yamal and Polar TransUrals. In: V. K. Ryabitsev (Editor), Materials to distribution of birds at Urals, TransUrals and Western Siberia. *Ekaterinburg*, UD of RAS Press, 94-116. (In Russian).
Khantemirov, R. M. (2000). 4309-year-old trees-ring chronology for Yamal and its use for reconstruction of a climate of the past in the north of Western Siberia. *Problems of ecological monitoring and modelling of ecosystems, Vol. XVII.*, 287-301. (In Russian).
Klausnitzer, B. (1987). Ökologie der Großstadtfauna. Jena: G. *Fischer*, 1-225.
Kosintsev P. A. (2006). Ecology of medieval people of the north of Western Siberia. Sources. *Ekaterinburg – Salekhard*, Ural State Univ. Press, 1-272. (In Russian).
Kotelina, N. S., Turubanova, L. V. & Teterjuk, B. U. (2000). Test of grasses. In: Restoration of the earths on the Far North. *Siktivkar*, 49-83. (In Russian).
Krivenko, V. G. (1991). *Waterfowls and its protection*. Moscow, 1-271. (In Russian).

Kryuchkov, V. V. (1976). *Sensitive Subarctic*. Moscow, Science Press, 1-136. (In Russian).

Kumari, E. V. (1957). Dynamics of Baltics ornitofauna for last centuries and probable directions of its further changes. *In: Questions of ecology: abstracts of third ecological conference*. Kiev, 277-284. (In Russian).

Kucheruk, V. V., Kovalevskiy, Ju.V. & Surbanos, A. G. (1975). Change of the population and fauna of birds in Southern Yamal for last 100 years. Bulletin of Moscow society of environmental researchers. *Dev. Biology, Vol. 80., 1,* 52-64. (In Russian).

Lamb, H. H. (1977). Climate: Present, past and future. Vol. 2. *Climate history and the future*. London: Methuen&Co Ltd, 1-835.

Leonovich, V. V. & Uspensky, S. M. (1965). Features of a climate and birds life in Arctic. *In: Proceedings of Institute of biology of UD of RAS*. Ecology of vertebrate animals of the Far North. Fascicle 38. Sverdlovsk, 141-148. (In Russian).

Magomedova, M. A, Morozova, L. M. (1997). A vegetative cover. In: L N. Dobrinskii (Editor), Monitoring of the biota of the Yamal peninsula in connection with development of objects for gas extraction and transportation. Ekaterinburg, URC "AeroCosmoEcology" Press, 11-99. (In Russian).

Magomedova, M. A., Morozova, L. M. & Ektova, S. N. et al. Yamal Peninsula: a vegetative cover. Ed. Gorchkovskiy P. L. Tyumen, City-press. 1-360. (In Russian).

Mazepa, V. S. (1999). By years reconstruction of average summer temperature of air in the north of Western Siberia since 1690 on the basis of the data about a radial gain of trees. *The Siberian ecological journal, 2*, 175-183. (In Russian).

Malchevskii, A. S. & Pukinsky, Ju. B. (1983). Birds of Leningrad region and adjacent territories: History, biology, protection. Vol.1. Leningrad, Leningradsky Univ. Press, 1-410. (In Russian).

McKendrick, J. D., Batzli, G. O., Everett, K. R. & Swanson, J. C. (1980). Some Effects of Mammalian Herbivores and Fertilization on Tundra Soils and Vegetation. Arctic and Alpine Research, *12*, 565-578.

Menzbir, M. A. (1882). *Ornithological geography of the European Russia. Part 1*. Moscow, Univ. Press, 1-524. (In Russian).

Mechnikova, S. A. & Kudryavtsev, N. V. (2005). Nesting of prey birds in forest-tundra of Southern Yamal in 2005. In: V. K. Ryabitsev (Editor), Materials to distribution of birds at Urals, TransUrals and Western Siberia. *Ekaterinburg*, UD of RAS Press, 204-209. (In Russian).

Mechnikova, S. A., Kudrjavtsev, N. V. & Luzan, P. I. (2005). The new data about distribution and number dynamics of some rare and not numerous birds in the south of Yamal. In: V. K. Ryabitsev (Editor), Materials to distribution of birds at Urals, TransUrals and Western Siberia. *Ekaterinburg*, UD of RAS Press, 209-212. (In Russian).

Mineev, Ju. N. (2003). Anseriformes of the East-European tundra. *Ekaterinburg*, UD of RAS Press, 1-225. (In Russian).

Moller, P. H. (1992). Decrease in number of Starlings (*Sturnus vulgaris*) wintering in the Leypark, Tilburg. Limosa, *65*, 19-22.

Morozov, V. V. (1995). Current status, distribution and population trends of Lesser White-fronted Goose (*Anser erythropus*) in Russia. *In: Bulletin of the goose study group of Eastern Europe and Northern Asia*, *1*, 131-144. (In Russian).

Morozov, V. V. (1999). Surveys for Lesser White-fronted Goose in the Bolshezemelskaya tundra, European Russia, in 1999. Fennoscandian Lesser White-fronted Goose conservation project: Annual report, 35-38.

Morozov, V. V. To fauna and distribution of birds in Bolshezemelskay tundra and on Yugor peninsula. In: V. K. Ryabitsev (Editor), Materials to distribution of birds at Urals, TransUrals and Western Siberia. *Ekaterinburg*, UD of RAS Press, 110-116. (In Russian).

Morozov, V. V. (1998). Pallid Harrier (*Circus macrourus*) in the south of Yamal. *Russian journal of Ornithology*, Sp. Is. *47*, 3-5. (In Russian).

Olofsson, J., Kitti, H., Rautiainen, P., Stark, S. & Oksanen, L. (2001). Effects of reindeer on composition of vegetation, productivity and nitrogen cycling. *Ecografy*, *24*, 13-24.

Olofsson, J. & Oksanen, L. (2002). Role of litter decomposition for the increased primary production in areas heavily grazed by reindeer, a litterbag experiment. *Oikos*, *96*, 507-515.

Paskhalny, S. P. (1985). To fauna of waders and passerines in the Arctic tundra of Yamal. *In: Distribution and number of vertebrate animals of Yamal*. Sverdlovsk, 34-38. (In Russian).

Paskhalny, S. P. (2004). Birds of anthropogenic habitats of Yamal peninsula and adjoining territories. *Ekaterinburg*, UD RAS Press, 1-166. (In Russian).

Paskhalny, S. P., Karagodin, I. J., Nesterov, E. V. & Golovatin, M. G. (1998). Nesting of Snow Bunting in anthropogenous habitats of Polar TransUrals. In: V. K. Ryabitsev (Editor), Materials to distribution of birds at Urals, TransUrals and Western Siberia. *Ekaterinburg*, UD of RAS Press, 129-130. (In Russian).

Paskhalny, S. P. & Sinitsyn, V. V. (1997). New data about rare and poorly known birds of Ob'River region and Polar Urals. In: V. K. Ryabitsev (Editor),

Materials to distribution of birds at Urals, TransUrals and Western Siberia. *Ekaterinburg*, UD of RAS Press, 119-122. (In Russian).
Peshkova, N. V. (1977). Productivity of vegetative communities at field station "Xadita" and influence of rodents on a grassy cover of polygons. *In: Biocoenotic role of animals in forest-tundra of Yamal.* Sverdlovsk, 134-145. (In Russian).
Plessky, P. V. (1948). Materials to ornitofauna of the Kirov region. *Scientific notes of Kirov GPU*, Is. 4. Kirov. 33-68. (In Russian).
Pugachuk, N. N. (1965). Waterfowls of Yamal peninsula. *In: Geography of waterfowl birds resources in the USSR. Part*, 2. Moscow. 57-58. (In Russian).
Rebristaya, O. V. (2006). Vascular plants. In: P. L. Gorchkovskiy (Editor), Yamal Peninsula: vegetative cover. Tyumen, City-press, 16-69. (In Russian).
Robertson, J. G. M. & Scoglund, T. (1985). A method for mapping birds of conservation interest over large areas. *In: Proceedings of VIII Int. Conf. on Bird Census and Atlas Work Birds census and atlas studies.* Tring: BTO. 67-72.
Robinson, R. A., Siriwardena, G. M. & Crick, H. Q. P. (2005). Status and population trends of Starling *Sturnus vulgaris* in Great Britain. *Bird Study*, *52*, 252-260.
Rubinshtejn, E. S., Polozova, L. G. (1966). Current climate change. *Leningrad*, 1-267. (In Russian).
Ryabitsev, V. K., Alekseeva, N. S., Polents, E. A. & Tyulkin, Ju. A. (1995). Avifauna finds on Middle Yamal. In: V. K. Ryabitsev (Editor), Materials to distribution of birds at Urals, TransUrals and Western Siberia. *Ekaterinburg*, UD of RAS Press, 64-66. (In Russian).
Ryabitsev, V. K. & Taylor, M. (2001). The plumage polymorphism and nomadism of Pomarine Skua *Stercorarius pomarinus* on the Yamal peninsula). *Russian Journal of Ornithology.* Express-issue, *140*, 307-313.
Sandford, S. (1976). Pastoralism under pressure. *ODI Rev.*, *2*, 45-68.
Shtro, V. G. (2003). Dynamics of rodent numbers on the Yamal peninsula and the impact on the Arctic fox population. Arctic Birds: *Newsletter of International Breeding Conditions Survey*, 5, 41-45.
Shtro, V. G. & Sokolov, A. A. (2006). To ornitofauna of Naduy-yaxa river basin, Middle Yamal. In: S. P. Paskhalny (Editor), The Scientific bulletin of the Yamal-Nenets autonomous region: Plant ecology and animals of the north of Western Siberia. *Salekhard*, Is. *6(2) (43)*, 61-65. (In Russian).
Shtro, V. G., Sosin, V. F. (2004). Some features of number dynamics of the Siberian lemming in subzons of tundra at Yamal. In: S. P. Paskhalny (Editor), The Scientific bulletin of the Yamal-Nenets autonomous region: Materials to

flora and fauna of Yamal-Nenets autonomous region. *Salekhard*, Is. *3(29)*, 110-115. (In Russian).

Shyatov, S. G., Mazepa, V. S. (1995). *Climate*. In: L. N. Dobrinskii (Editor), Nature of Yamal Ekaterinburg: Science Press. of UD of RAS, 32-68. (In Russian).

Shyatov, S. G. & Mazepa, V. S. (2007). Climatogenic dynamics of forest-tundra vegetation in Polar Urals. *Lesovedenie*, 6, 11-22. (In Russian).

Shukhov, I. N. (1915). Birds of Obdorsky region. *YB of zoological museum of academy of science*, *Vol. 20*, 167-237. (In Russian).

Sokolov, V. A. & Sokolov, A. A. (2005). Interesting meetings of birds in the southwest of Yamal in 2005. In: V. K. Ryabitsev (Editor), Materials to distribution of birds at Urals, TransUrals and Western Siberia. *Ekaterinburg*, UD of RAS Press, 243-246. (In Russian).

Sokolov, V. A., Sokolov, A. A., Fisher, S. V. & Ogarkov, A. E. (2001). New data about distribution of birds in the southwest of Yamal. In: V. K. Ryabitsev (Editor), Materials to distribution of birds at Urals, TransUrals and Western Siberia. *Ekaterinburg*, UD of RAS Press, 144-147.(In Russian).

Sokolov, V. A., Kornev, S. V., Sokolov, A. A. & Ogarkov, A. E. (2002). New data about not numerous, rare and protected birds on southwest Yamal. In: V. K. Ryabitsev (Editor), Materials to distribution of birds at Urals, TransUrals and Western Siberia. *Ekaterinburg*, UD of RAS Press, 237-239. (In Russian).

Sotnikov, V. N. (2006). Birds of the Kirov region and adjacent territories. Vol. 2. Passeriformes. Part 1. *Kirov*, 1-448. (In Russian).

Syroechkovskiy, E. E. (1995). Changes in nested distribution and number of Red-breasted Goose in 1980-1990th years. *Bulletin of the goose study group of Eastern Europe and Northern Asia. Moscow*. *1*, 89-102. (In Russian).

Tarasov, V. (1915). Report about trip on Yamal peninsula with S.I. Drachinsky's veterinary expedition in 1913. In: YB of the Tobolsk museum. Is. 24. *Tobolsk*, 1-32. (In Russian).

Yerokhov, S. N. (2003). Graylag goose in Kazakhstan: distribution, numbers, and the main stages of the annual cycle. Cazarka: Bull. of the goose, *swan and duck study group of Northern Eurasia*. Moscow. *9*, 103-135. (In Russian).

Zimov, S.A., Chuprynin, V. I., Oreshko, A. P., Chapin, III F. S., Reynolds, J. F. & Chapin, M. C. (1995). Steppe-tundra transition: A herbivore-driven biome shift at the end of the Pleistocene. *American Naturalist*, *146(5)*, 765-794.

Interactive Agricultural Ecological Atlas of Russia and Neighboring Countries: Economic Plants and their Diseases, *Pest and Weeds*, *2008*, http://www.agroatlas.ru/en/

In: Tundras: Vegetation, Wildlife...
Editors: B. Gutierrez et al. pp. 47-79

ISBN: 978-1-60876-588-1
© 2010 Nova Science Publishers, Inc.

Chapter 2

THE RECIPROCAL RELATIONSHIPS BETWEEN HIGH LATITUDE CLIMATE CHANGES AND THE ECOLOGY OF TERRESTRIAL MICROBIOTA: EMERGING THEORIES, MODELS, AND EMPIRICAL EVIDENCE, ESPECIALLY RELATED TO GLOBAL WARMING

O. Roger Anderson
Biology, Lamont-Doherty Earth Observatory of Columbia University, Palisades, NY 10964 U.S.A.

ABSTRACT

High-latitude, moss-rich tundra communities (e.g., *Sphagnum* and *Hylocomium* spp.) are circumpolar in distribution, including conifer forests and tundra ecosystems that occupy millions of square kilometers. The sheer geographic scale of these high latitude biomes is sufficient to warrant scientific interest. However, it is becoming increasingly clear that major changes in high latitude climate patterns may have significant affects on the ecology of these communities. In turn, changes in the life histories, physiology, and productivity of the biota may also directly, or indirectly, influence local to global climate patterns; especially the balance of atmospheric carbon dioxide that is sequestered by primary production versus that released by respiratory activity – thus, potentially influencing global warming. Substantial attention has been given to aboveground biota, particularly the role of plants in this biotic-climatic reciprocal relationship, notably in relation to global warming and likely changes in annual mean temperature and precipitation patterns across vast geographic regimes at high latitudes. However, belowground processes also are likely to be substantially affected, especially the response of microbiota. Changes in the biology of terrestrial microbial communities may be directly affected by local meteorological factors, but also indirectly by effects of above- and belowground coupling. This coupling includes the effects of climate variables on plant physiology, especially the degree of primary productivity, release of organic compounds into the soil and their influences on the productivity and respiratory activity of associated belowground microbiota (e.g., bacteria, fungi and eukaryotic microbes, including protozoa). The major groups of protozoa include heterotrophic flagellates, naked amoebae (lacking a shell) and testate amoebae (enclosed in an organic or mineralized shell). With increasing evidence that the tundra permafrost is incurring prolonged seasonal warming and thawing to greater depths, there is an increased probability that associated microbial communities, that are normally more dormant during much of each annual climatic cycle, may become increasingly metabolically active. Given the enormous stores of plant-derived organic matter that have accumulated and remained frozen during millennia, there is substantial potential for enhanced terrestrial microbial respiration and significant release of atmospheric carbon dioxide. Nearly one-third of the global terrestrial carbon is stored in these high latitude environments. Currently, there is increasing interest in the complexities of the responses of terrestrial microbial communities to high-latitude climatic changes and the likelihood that they could have a significant effect on global warming through elevated respiratory activity. Some of the current emerging theories, models, and recent empirical evidence for the dynamics of these reciprocal interactions between climate and terrestrial microbial communities are reviewed, with particular attention to biogeochemical and ecological perspectives.

INTRODUCTION

Tundra ecosystems, lying largely along the Arctic Circle, reach as far south as 60° N at the neck of the Kamchatka peninsula and include southern extensions into the Scandinavian, Timan and Ural mountains. The moss-rich, high latitude tundra and boreal forests cover millions of square kilometers worldwide (O'Neill, 2000; Walker and Walker, 1996) and represent a major reservoir of organic carbon. Currently, many tundra ecosystems are net carbon sinks annually due to plant primary production, but with increasing global warming, and likely melting of the permafrost, there is concern that these vast regions may become a net carbon source through respiratory loss of CO_2, contributing to increased atmospheric carbon dioxide concentration. For over 40 years, the composition, biogeochemistry and ecology of Alaskan and high latitude soils have been the subject of extensive research (e.g., Bockheim, et al., 1999) yielding a substantial database to support more robust studies of tundra ecology.

Moreover, during the last quarter century, the effects of climate, especially global warming, on the ecology of above- and belowground processes have received increasing attention (e.g., Schröter, et al., 2004; Wardle, et al., 2004; Wolters, et al., 2000). Among the issues addressed are to what extent the tundra, exemplified by its particular history, relatively harsh climate for life, and the persistence of permafrost throughout recent Holocene history is similar to other terrestrial environments and in what ways is it unique, requiring special consideration in making generalizations about future forecasts as we envision an increasingly warmer planet (e.g., Broecker, 1975).

Approximately one third of global carbon is sequestered in the permafrost soils of the tundra; it is estimated that it represents two-thirds of the total atmospheric carbon as carbon dioxide and as much as 11% of terrestrial soil organic matter (Grogan and Jonasson, 2005; Schimel and Mikan, 2005). The fate of this stored carbon, both presently and in the future, if global warming trends continue, is an important consideration ecologically and also in terms of climate dynamics; because the carbon metabolized as CO_2 by terrestrial microbiota may contribute substantially to a feed-forward effect exacerbating the greenhouse effect and leading to a spiraling upward of global warming (e.g., Boddy, et al., 2008).

The tundra, while distinctive in its high latitude seasonal cycles and uniquely adapted biota, is ecologically similar to other terrestrial ecosystems with respect to modeling the broad categorical components of this biome (e.g., Figure 1). In general, as diagrammed in Figure 1, above ground processes include the abiotic and biotic factors that affect ecosystem functioning, such as climate and related variables and the role of life forms (plants and animals) as they interact with their natural environment and with one another. The products of these above ground processes also affect the quality of the environment and life processes below ground. Temperature of the atmosphere and radiant solar energy affect soil temperature, and particularly in high latitudes the thaw depth of the permafrost. Precipitation provides a major source of soil moisture and in optimal amounts (less than at anaerobic saturation) is a major source of water essential for life, especially the microbiota. Abiotic and biotic sources of CO_2 provide a carbon source for plant primary production (including vascular and nonvascular plants).

Leaf fall, litter and other sources of plant biomass above ground contribute to organic matter that enters the soil. Roots and their organic exudates, resulting from aboveground photosynthesis, are a major source of solid and dissolved organic matter that supports microbial metabolism. This is augmented by organic wastes from animals, and their remains, that provide additional carbon resources for heterotrophic microbes.

The terrestrial microbial community below ground is a diverse assemblage of organisms in three major biological kingdoms: Monera (bacteria), Fungi, and Protista (largely amoebae and flagellates, but also including ciliates in lower numbers and often more frequently encysted in soils). Furthermore, photosynthetic algal protists and cyanobacteria on the soil surface provide additional sources of primary fixed carbon and nitrogen. In addition to the uptake of atmospheric CO_2 by the plant community (particularly, mosses and scattered vascular plants in the tundra); respiratory CO_2 released by plant roots, microfauna, and especially the robust microbial community eventually diffuses into the atmosphere and can represent a major contribution to global atmospheric CO_2. In turn, this source of CO_2 can drive higher rates of primary production. However, if the net efflux exceeds the amount fixed by primary producers, the increasing load of atmospheric CO_2 can substantially increase the greenhouse effect promoting increased global warming. This complex set of abiotic factors is represented by white arrows, and all of the biotic factors in the carbon flux are represented by opaque arrows (Figure 1). The magnitude and rates of these processes are of substantial importance in understanding the dynamics of aboveground and belowground coupling of biogeochemical processes. Although this box model is somewhat of an over simplified view of the complexities of the processes, it provides a general perspective for the organization of the chapter.

This chapter addresses current understanding of how the tundra microbiota may be influenced by increasing temperatures and other climatic variables at high latitudes; including, microbial community composition, and metabolic activity, and their mineralization of stored carbon sources, releasing respiratory CO_2, with consequent changes in temperature and its influences on climate dynamics, locally and globally. Hence, the literature reviewed is limited to, and largely focused on, relatively recent research pertinent to this theme. Methane emissions from methanogens are also a significant contributor to global warming and additional tundra research attention is warranted (e.g. Kotsyurbenko, et al., 1996; Nozhevnikova, et al. 2000). However, to maintain a reasonable length and a consistent focus, only the role of respiratory CO_2 will be considered in this chapter. The theme is addressed in two major sections: 1. an overview of tundra carbon cycle dynamics and its relationship to reciprocal relations between aboveground and belowground processes using the box model (Figure 1) as a context, especially emphasizing likely interactive effects of climate with the subsurface microbial community; and 2. a review of some current and emerging models relevant to the theme of the role of microbial communities in the carbon cycle and global warming, especially in relation to climatic and other environmental variables.

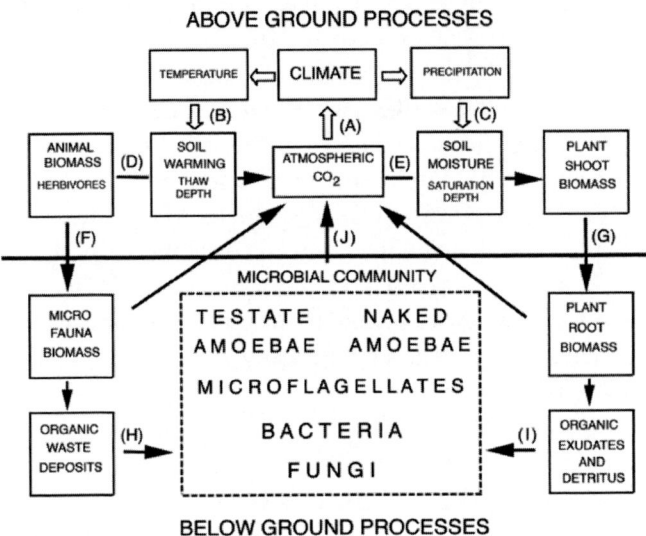

Figure 1. Coupling of above- and belowground processes. Above ground: Atmospheric CO_2 concentrations (A), when increasing, can elevate global temperature (B) and permafrost thaw at higher latitudes as well as changes in amount and geographic distribution of precipitation (C) influencing water table height and soil saturation leading in excess to anaerobiosis and depressed respiratory CO_2 emission, or in moderate amounts to increased respiration. Respiration by animals (D) adds to atmospheric CO_2 that in turn can be assimilated by plants (E) during photosynthesis. Animal organic waste products (F) and plant litter (G) contribute to the belowground pools of carbon resources. Below ground: Organic waste deposits from microfauna (H) and from plant root biomass and its exudates (I) are sources of mineral and carbon nutrients for the microbial community, especially through decay processes mediated by fungi and some bacteria. The respiratory CO_2 flux (J) from the microbes and also from microfauna and plant roots, if enhanced by warmer temperatures, contributes to additional atmospheric CO_2, further increasing possible warming through an elevated greenhouse effect.

GENERAL PERSPECTIVES AND RELATED DETAILED INFORMATION

A broad range of high latitude studies have addressed critical variables influencing CO_2 exchange with the atmosphere, especially in relation to coupling of above- and belowground processes, including 1. monitoring carbon turnover related to climatic and environmental factors across geographical regions (e.g., McMichael, et al., 1999; Shaver, et al., 2006), 2. effects of elevated temperature on respiratory loss of CO_2 (e.g., Hobbie and Chapin, 1998; Mertens, et al., 2001; Shaver and Jonasson, 1999; Welker, Brown and Fahnestock, 1999), 3. variation in CO_2 exchange with the atmosphere due to differences in site-based meteorological and edaphic factors (e.g., Eugester and Chapin, 2005; Harding, et al. 2002; Kutzbach, 2006; Sommerkorn, 1998; Yoshimoto, Harazono and Oechel, 1997), and 4. responses to seasonal and inter-annual variability (e.g., Lafleur, Griffis and Rouse, 2001; Mano, et al., 2003; Oechel, et al., 1995). Some representative research findings are reviewed here, first in relation to aboveground factors and secondly as related to belowground factors that especially influence microbial respiratory CO_2 flux.

Aboveground Abiotic and Biotic Factors

In the tundra ecosystem, above ground environmental changes associated with global warming may have significant effects on endogenous biota, including changes in the range and density of rodents and other small herbivores consuming plant biomass (e.g., Callaghan, et al., 2004) that may affect primary production and the balance between CO_2 sources and sinks (Sjogersten, van der Wal, and Woodin, 2008). These perturbations include alterations in soil invertebrate communities (e.g., Coulson, et al., 1996; Couteaux and Bolger, 2000), changes in shrub cover and herbaceous plant distribution (Chapin, et al., 1995; Walker, et al., 2006), and possible northward expansion of the boreal forests, encroaching on the steppe (Calef, et al., 2005). Presently, however, current evidence from satellite imagery suggests that this advance is not yet substantial (Masek, 2001). Moreover, there may be considerable variability in changes in vegetation community structure, including advances or retrenchment of boreal forests, depending on local climate and geographic features (e.g., Skre, et al., 2002).

Currently, there is substantial variability in tundra vegetation communities depending on geographic location, slope, elevation, and soil moisture, especially along gradients from high water table saturated soils to more mesic and dry cold desert sites (e.g., McFadden, Eugster and Chapin, 2003; Mcguire, et al., 2002; Meyer, et al., 2006; Ping, Michaelson and Kimble, 1997; Schmidt and Boelter, 2002; Williams and Rastetter, 1999). However, given the close coupling of belowground processes with the overlying faunal and especially plant communities, further research is needed to fully understand how the belowground microbial community may respond to climatic changes driving alterations in aboveground tundra plant ecology.

There is good evidence, for example, from field studies and laboratory investigations, that increasing atmospheric CO_2 affects plant primary production, including increased release of root exudates enhancing microbial productivity and community structure (e.g., Anderson and Griffin, 2001; Drigo, Kowalchuk and van Veen, 2008) and increased fine root production by forest trees (e.g., Lichter, et al., 2005) that may affect rhizosphere microbial dynamics. Further research at high latitudes is needed to more fully document how tundra plant and microbial communities will respond to elevated atmospheric CO_2, especially in relation to changing microbial community dynamics, productivity, and possible enhanced respiratory CO_2 release (e.g., Oberbauer, Oechel and Riechers, 1986; Oechel, et al., 1994). Particular attention is needed to assess additional climatic influences on microbial communities beyond the effects of likely elevated temperature resulting from global warming (e.g., Anderson, 2008; Loya and Grogan, 2004); especially, given the evidence that arctic tundra soils are unique in many ways and possess greater bacterial diversity than other high latitude ecosystems such as boreal forests (Neufeld and Mohn, 2005).

Belowground Abiotic and Biotic Factors

Nitrogen nutrient sources. Changes in the aboveground climate, including variations in precipitation and temperature, contribute to variations in belowground microbial community responses, due in part to infiltration of nutrients, changes in plant productivity, and the release of photosynthetic organic products from roots to the soil. Increased precipitation may enhance percolation of mineral as well as organic nutrients from aboveground into subsurface soil. Inorganic and organic nitrogen compounds, in addition to organic carbon and mineral nutrients, are considered to be major sources of nutrition for microbial communities and essential metabolites for respiratory CO_2 production (e.g., Buckeridge and Grogan, 2008; Kaiser, et al., 2005; Nordin, Schmidt and Shaver, 2004; Schmidt, et al., 2002; Sorensen, Michelsen and Jonasson, 2008). Moreover, there is considerable evidence that rhizosphere microbes, in lower latitudes can contribute significantly to mineralization and recycling of soil nutrients enhancing fertility of the soil and promoting plant growth. However, there is a growing interest in elucidating microbial/plant nutrient relations at higher latitudes. For example, Schimel and Chapin (1996) have shown that some arctic sedges (e.g., *Eriophorum vaginatum* and *Carex aquatilis*) are capable of assimilating amino acid and ammonium nitrogen in situ, and may compete effectively with microbes for these nitrogen sources. However, it is important to recognize that the microbial community serves a significant role in the degradation of larger biomolecules and release of amino acid residues and other smaller organic molecules that may become accessible to plants. Further studies on the complexities of the dynamics of nutrient cycling in high latitudes, including the sources and sinks of terrestrial nutrients, are warranted.

With anticipated increases in nitrogen supply in tundra ecosystems due to global warming, Rien (2009) examined the amount of nitrogen and phosphorous transferred to arctic soil in plant litter when increased nitrogen supply was experimentally added to the soil system. He found plant species-specific variations; but overall, the nitrogen amendment increased the amount of nitrogen and phosphate in dead leaves, and hence the amount of organic nitrogen and phosphate transferred to the soil due to input from leaf litter. Given that nitrogen is often a limiting nutrient in some soil systems, considerable attention has been devoted to examining the role of nitrogen fertilization as well as increased atmospheric CO_2 on soil microbial activity and respiration. These studies show that the response of Arctic tundra soil microbial communities to elevated nitrogenous compounds is complex, with evidence that nitrogen fertilization may suppress some microbial functional groups resulting in a depression of overall microbial biomass and activity in tundra soil (Schmidt, et al., 2004).

In other cases, depending on locale and the chemical composition of the amended nitrogen (e.g., ammonium vs. nitrate), there may be a significant effect on bacterial production (assessed by ^{3}H thymidine incorporation into DNA), but not necessarily an increase in total bacterial biomass (Stapleton, et al., 2005). However as Stapleton, et al. noted, the complexity of the microbial food web should be taken into account when interpreting these data. If carbon nutrients are limiting, then any potential enhancement effect of nitrogen amendation on bacterial growth could be masked by top-down predatory pressures from eukaryotic microbes (e.g., amoebae and flagellates) or some invertebrates (e.g., nematodes). As elsewhere in terrestrial communities, their data showed that the main bacterial predators were heterotrophic flagellates and naked amoebae, which also evidenced no significant response to the nitrogen addition. The flagellates, however, showed a correlation with bacterial abundance suggesting a dependence on bacteria as a food source. Relatively little is known about terrestrial microbial food webs at high latitudes; but in general, we know that bacteria are at the base of food webs in most terrestrial environments and are a direct prey for heterotrophic flagellates. Amoeboid protists and ciliates also prey on bacteria. In addition some amoebae consume smaller flagellates and occasionally small ciliates. Some research attention has been given to high latitude aquatic microbial food webs indicating that they possess similar trophic levels (although sometimes less complex) as described for soil environments (e.g., Hobbie and Rublee, 1999; Sawstrom, et al., 2002). Overall, it appears that the chemical form of nitrogen compounds and their concentration, in addition to the locale and composition of the soil, may be important factors in determining the response of tundra microbial communities; in some cases promoting productivity or suppressing one or more of the constituent populations depending on the dynamics of their trophic interactions and effects of abiotic environmental variables.

Organic carbon nutrient sources. Carbon content of tundra soils varies substantially by locale and edaphic conditions of the site. For example, substantial differences are found in the trough, rim and bare soil of frost-boil tundra, with 23.9 kg C m^{-2} in the troughs, while drier rims contained only 50% and bare soil patches about 17% of the carbon found in troughs (Kaiser, et al., 2005). There were similar results for nitrogen. Soluble organic carbon (SOC) content was examined in nine geographic area units at Barrow, Alaska (Bockheim, et al., 1999). SOC varied from 2.5 kg m^{-3} in some modern beach sediments to > 73 kg m^{-3} in Typic Sapristels in high-centered, ice-wedge polygons developed in reworked organic-rich lake sediments. The average SOC for the entire 64-km^2 terrestrial area was c. 50 kg m^{-3}. The variation in SOC within soil map units and individual patterned-ground units was attributed primarily to differences in the amount of ground ice.

Gundelwein, et al. (2007) studying Arctic tundra in Siberia, estimated that the mean stock of carbon was 14.5 kg m^{-2} within the active layer, and a total of about 31 kg C m^{-3} in the entire upper meter of the soils. Further studies in Alaskan, Arctic pedons by Ping, et al. (1997) reported a range of 169 Mg C ha^{-1} to 1,292 Mg C ha^{-1}. Carbon storage in an arctic coastal marsh pedon amounted to 692 Mg C ha^{-1} and for an arctic tundra pedon, a range of 314 to 599 Mg C ha^{-1}. A lower amount was detected in mixed forest and coastal forest pedons, i.e., 240 and 437 Mg C ha^{-1}, respectively.

Given the importance of phytomass contributions to high latitude soil carbon stocks, Kolchugina and Vinson (1993) analyzed the component content in permafrost in Russia and reported that under the present climate, the phytomass (live vegetation, above- and belowground) carbon pool was 17.0 Gt; whereas, the mortmass (coarse woody debris) carbon pool was 16.1 Gt. The litter carbon pool was 6.4 Gt C and the soil carbon pool including peatlands was 139.4 Gt. Moreover, live vegetation and plant detritus (mortmass and litter) taken together were approximately one-third of the soil carbon pool. In addition to soluble (SOM) and particulate (POM) organic matter derived from vascular plants, additional contributions can come from the dead and dying remains of moss (i.e., *Hylacomium* and *Sphagnum*) including substantial pulse release of SOM from living surface-dwelling moss during periodic rain storms (e.g., Wilson and Coxson, 1999).

Carbon sources vary in chemical complexity and in refractory quality, with small molecular weight compounds (e.g., amino acids and sugars) providing more immediately available nutritional resources, while more complex polymerized compounds such as cellulose, lignin and humic substances require a longer time to be degraded and made available to bacteria and other microbes at the base of the food web. The kinetics of decomposition can be represented by a simple first-order equation (eq. 1) with a series of terms representing each of the biochemical groups in the terrestrial substrate:

$$Y_t \text{ (Biomass residue)} = C_1 \cdot \exp(-k_1 t) + C_2 \cdot \exp(-k_2 t) \qquad (1)$$

where C_1 and C_2 are concentrations of the individual organic carbon fractions, t is time, and k_1 and k_2 are decomposition rate constants specific to each organic compound, with smaller k values for more refractory constituents. Additional terms can be added to the equation to represent other refractory substances, but usually a biphasic or triphasic equation with two or three terms, respectively, is sufficient (e.g., Aber, et al., 1990; Tate, 1995; Trofymow, Preston and Prescott, 1995). Additional details of some relevant models describing the transformations of these carbon sources in soil microbial communities are presented below in the section on Current and Emerging Models.

Soil microbial populations. In addition to the available nutritional resources driving microbial metabolism and CO_2 respiratory production, the population composition and dynamics of the belowground microbial communities are also of critical importance in fully accounting for belowground sources of atmospheric CO_2. Within the limited scope permitted in this review, some data on the relative abundances of tundra-dwelling bacteria, fungi, and protists are presented. Fungal and bacterial biomass and biovolume were estimated by Schmidt and Boelter (2002) in an arctic transect in northern central Siberia from 11 permafrost soil pits. They report a fungal biovolume of up to 3.5 mm^3 g^{-1} dry wt. (median 0.19 mm^3 g^{-1} dry wt.) and maximum hyphal length of 393 m g^{-1} dry wt. (median 21 m g^{-1} dry wt.). Fungal biomass was found up to 455 µg C g^{-1} dry wt. (median 24 µg C g^{-1} dry wt.). Moreover, bacterial counts ranged from 0.16 to 7.38 x10^9 g^{-1} dry wt. and bacterial biomass was much smaller than for fungi and ranged from 0.68 to 20.38 µg C g^{-1} dry wt. (median 6.19 µg C g^{-1} dry wt.), much less than fungi because of their small cell volume (median 0.04 µm^3). The median ratio of fungal to bacterial biomass was 4.5, with a broad range of 0 to 174.1, indicating that fungi in general dominated the biomass.

However, variations in relative abundance were found across a tundra landscape (upland tussock, stream-side birch-willow, and lakeside wet sedge tundra) in Arctic Alaska, USA (Zak and Kling, 2006). They report that microbial community composition and function were distinct among tundra ecosystems, with tussock tundra containing a significantly greater abundance and activity of soil fungi. However, variations in relative abundance of fungi and other microbiota also may occur with soil depth and across seasons as reported for some boreal forest sites. The depth distribution of bacteria, fungi and protozoa was examined at three Danish forest sites situated on a Weichel moraine (Elkelund, Ronn and Christensen, 2001). Except for a bacterial peak at 42.5 cm in the peat profile, the general tendency for all groups was a decrease in biomass with increasing depth. Protozoa decreased more rapidly with increasing depth than the other two groups of organisms examined. However, it is important to note that an evaluation of the bacterial-protozoan relationship using a simple mathematical model indicated that the subsurface protozoan populations are active and not accidental percolated cysts.

In further studies in northern Sweden, Bjork, et al. (2008) found two seasonal shifts in microbial composition in forest soil. The first was associated with snowmelt accompanied by a decrease in fungal marker fatty acids and an increase in Gram-positive- and actinobacterial marker fatty acids, resulting in a decreased ratio of fungi-to-bacteria. The second shift occurred across the growing season with an increase in markers for Gram-negative bacteria. Further evidence of the relative contribution of fungi and bacteria to respiratory CO_2 efflux was obtained for tundra and boreal forests by Ananyeva, et al. (2006) using a technique of selective inhibition of substrate induced soil respiration. At these sites, fungi are major contributors to the efflux (63-82%). The fungal-bacterial ratio in tundra soils (0-5 cm without litter) was 4.3, which is consistent with median ratios cited above by Schmidt and Boelter (2002).

Relative abundances of bacteria and protists (naked amoebae, testate amoebae and heterotrophic nanoflagellates) in tundra soil during spring and summer at

Toolik, Alaska were reported by Anderson (2008), including the contribution of each taxon to calculated respiratory CO_2 efflux (Table 1). This complete account of the carbon content and respiratory CO_2 in each of the major groups of protozoa was not possible previously due to inadequate methods of accounting for the contribution by naked amoebae, but is now possible by applying a recent method that permits estimations based on direct observation of living amoeba cell size (Anderson, 2006a). In both June and July samples, bacteria were the most abundant followed by heterotrophic small flagellates, and last by naked and testate amoebae. The same pattern was observed for the carbon content (μg C g^{-1} soil dry weight); but it is worth noting that the combined carbon content of the naked and testate amoebae in the June and August samples was substantial relative to the flagellate content (71% and 78%, respectively), with the testate amoebae accounting for the majority of the amoeba carbon in August. However, bacteria and heterotrophic flagellates accounted for the bulk of released respiratory CO_2. Heterotrophic eukaryotic microbes are abundant in many aquatic and terrestrial environments and can account for a substantial amount of the total carbon in microbial communities (e.g., Anderson, 2006b, 2007, 2008). The fate of this carbon as it moves through the food web and its ultimate release as respiratory CO_2 deserves additional research attention. In overview, based on all of the above-cited biomass measurements, the relative proportions of members of the tundra microbial community are fungi > bacteria > heterotrophic flagellates > amoebae. However, the relative abundances of each group may vary substantially across geographic sites especially in relation to type of vegetation, quality of soil and its nutrients, hydrology, and dynamics of interactions among the members of the microbiota within the soil community.

Given that testate amoebae are significant predators on soil bacteria and constitute a significant proportion of the amoeba populations in high latitude ecosystems, Beyens, et al. (2009) examined the potential effects of global warming on the structure of testate communities by experimentally simulating a heatwave in Greenland arctic soils. They found that while the experimental heating of the soil was sufficiently severe to induce significant leaf mortality in the aboveground vegetation, overall there was no detectable effect on testate amoebae abundance. However, transient shifts in species populations occurred in the heated plots during the exposure, followed by increases in species richness weeks after the experimental heatwave had ended. The most prominent testate taxa appearing after the heating event were bacterivorous genera in agreement with observational evidence of a transient peak in bacterial abundances caused by the heatwave. Lobose pseudopod-bearing testate amoebae were more resistant to the heating and its associated desiccation than filose amoebae.

Respiratory CO_2 flux to the atmosphere. With increasing evidence of global warming, considerable attention has been given to total CO_2 efflux from high latitude soils, either natural or artificially heated to simulate warming. Some examples of these data are presented here, in part as a context for the review of current and emerging models presented in the next section. Jones, et al. (1998) measured net CO_2 flux from moist tussock and dry heath at Toolik, Alaska for ambient and warmed (2 °C) plots. They estimated that during the snow-free season, unwarmed plot fluxes ranged from 11.5 to 33.2, and 38.2 to 40.3 g CO_2-C m^{-2} for dry and moist sites, respectively. At the warmed plots, however, they estimated ranges of 41.7 to 49.0, and 55.7 to 80.8 g CO_2-C m^{-2} were lost at the dry and moist sites, respectively. Overall, at these sites, Jones, et al. estimated that the net annual loss to the atmosphere ranged from 20 g CO_2-C m^{-2} to 500 g CO_2-C m^{-2}, a significantly different conclusion from results at other locations showing that the arctic tundra is a net CO_2 sink. Nobriega and Grogan (2008) measured soil respiration and net ecosystem carbon exchange (NEE) at three sites varying in moisture content in the Canadian low arctic tundra (dry heath, mesic birch, and wet sedge). Based on an analysis of their graphical data, the soil respiration at the three sites was respectively, c. 100, 175, and 50 g CO_2-C m^2. The NEE for the three sites was respectively, near zero, -30, and -80 g CO_2-C m^{-2}, indicating that during the snow-free season, the latter two were a net sink for CO_2 under existing ambient conditions. Further evidence of the complexities of long term climatic changes on seasonal soil respiratory flux in Alaskan Arctic (Oechel, et al., 2000) indicates that some ecosystems may have the capacity to metabolically adjust to long-term (decadal or longer) changes in climate, including changes in nutrient cycling, physiological acclimation, and population and community reorganization, resulting in less respiratory efflux with climatic warming. Nonetheless, despite the reported acclimation, the Arctic ecosystems that they studied were still annual net sources of CO_2 to the atmosphere (c. 40 g C m^{-2} yr^{-1}) due to winter release of CO_2. In general, there is growing interest in the role of snow cover and changing temperature patterns on winter release of soil respiratory CO_2. Oechel, et al. (1997) reported that significant amounts of soil carbon on the North slope of Alaska may be lost as CO_2 to the atmosphere during the fall, winter and spring months, with greatest losses during October and May when soil temperatures were at a maximum, and portions of the soil profile were unfrozen. The daily loss rate equated to a seasonal loss of approximately 70 and 20 g C m^{-2} $season^{-1}$ for tussock and wet sedge tundra sites, respectively.

CO_2 efflux and variation in soil environmental characteristics were examined by Oberbauer, Tenhunen and Reynolds (1991) in two tundra vegetation communities; i.e., water track (a small drainage of intermittent water flow) and tussock tundra, in the northern foothills of the Philip Smith Mountains in arctic Alaska. They found that during four of the six measurement periods, the rate of CO_2 efflux differed significantly between sites. Early in the season, respiration was greater in tussock tundra than at the water track, but later in the season, rates at the water track exceeded those at the tussock site. Efflux of CO_2 at both sites was positively correlated with soil temperature. However, soil surface (0 - 5 cm depth) environmental conditions were better predictors of CO_2 flux than were conditions measured at greater depth (5 -10 cm). The presence of soil moisture appeared to increase respiration between 100 and 700% of soil dry weight. However, there was a decrease in soil respiration at higher water contents, perhaps due to swamping and anaerobic conditions. In general, there is increasing evidence that water table is one of the primary controls on respiratory CO_2 loss to the atmosphere in wet tundra and bog systems (e.g., Billings, et al. 1982, 1983; Luken and Billings, 1985; Oberbauer, et al. 1992). The effects of soil moisture were stronger in tussock tundra than in the water track community. Overall, Tenhunen and Reynolds data suggest that both soil temperature and soil moisture limit CO_2 efflux in water track and tussock tundra communities and that the relative importance of these factors changes throughout the growing season. Similar site-specific variations in terrestrial respiratory CO_2 emissions at tundra locations on the Taimyr Peninsula (Siberia) were reported by Sommerkorn (1998) who found lower fluxes for depressions (e.g., 2.6 – 4.3 g m^{-2} day^{-1}) compared to those from higher elevations on tussocks (e.g., 4.7 – 10.9 g m^{-2} day^{-1}). The highest value, however, was recorded during an unusually warm sampling season.

Further research by Oberbauer, et al. (1992) on CO_2 efflux from riparian tundra in the Brooks Range, Alaska showed that as the growing season progressed, rainfall was low and furthermore depth to water table and soil temperature increased. As a response, CO_2 efflux increased markedly, attaining rates late in the season of approximately 10 g CO_2 m^{-2} day^{-1}. In addition, they made laboratory measurements of soil respiration as a function of temperature and water content. Their results indicated that the effect of these factors on microbial respiration may explain a large part of the diurnal and seasonal variation observed in CO_2 flux.

A meta-analysis of a range of experimental treatments investigating artificial heating of 32 varied sites representing four broadly defined biomes including high latitude tundra, low tundra, grassland and forest, showed that across all sites and years, 2-9 years of experimental warming significantly increased soil respiration rates by 20% and plant productivity by 19% (Rustad, et al., 2001). Moreover, the response of soil respiration to warming was generally larger in forested ecosystems compared to low tundra and grassland ecosystems, and the response of plant productivity was generally larger in low tundra ecosystems than in forest and grassland ecosystems. Furthermore, Oechel, et al. (1995) reported a significant difference in net ecosystem carbon balance of wet sedge sites in Barrow, Alaska for CO_2 flux measurements obtained in the 1991-1992 growing season compared to measurements in 1971. In their more recent 1990s measurements, high-center polygons were net sources of atmospheric CO_2 of approximately 14 g C m^{-2} yr^{-1} compared to low-center polygons that were losing only approximately 3.6 g C m^{-2} yr^{-1}. Moreover, on average, moist meadow habitats were also sources of approximately 1.3 g C m^{-2} yr^{-1}. However, ice wedge habitats were accumulating 4.0 g C m^{-2} yr^{-1}. They concluded that this difference in ecosystem function over the period of two decades may be due to an increase in surface temperatures in recent decades resulting in decreases in the soil moisture status.

Overall, these data point to the varying effects of climatic and geographic variables on respiratory CO_2 fluxes to the atmosphere, and suggest that with increasingly warmer and somewhat drier conditions, some tundra ecosystems may become net sources of atmospheric CO_2, especially if water table depth in some saturated soils increases, making the soil more aerated, thus supporting aerobic respiration. However, the complexities of differences in geography, and the amount and kind of vegetative cover, must also be taken into consideration when extrapolating results from specific locales to larger scale domains (e.g., Rustad, et al., 2001; Williams and Rastetter, 1999). Nonetheless, some progress is being made in estimating large-scale, global terrestrial respiratory CO_2 efflux accompanying predicted global warming (e.g., Raich and Schlesinger, 1992).

In summary, the overall data indicate that in most tundra sites, there is a net annual sink for atmospheric CO_2, varying particularly with the amount of vegetative cover, while some specific geographic locales may already be net CO_2 sources. During the plant dormant seasons from late summer through winter into early spring, there may be broadly a net respiratory flux of CO_2 into the atmosphere, including a low, but detectable, flux during snow cover in some locations. This CO_2 flux may be attributed to metabolism of more recalcitrant, residual phytomass (e.g., cellulose and lignin-rich litter) that was not mineralized during the plant growing season.

SOME CURRENT AND EMERGING MODELS

One of the more interesting, although challenging, aspects of research at high latitudes is modeling of belowground processes contributing to microbial community activity and flux of respiratory CO_2 to the atmosphere; especially as it applies to predicting effects resulting from, and contributing to, global warming. A variety of modeling studies have been done encompassing nutrient sources and flow, especially organics, within soil microbial communities. Other studies examined the value of predictive models relating environmental variables to microbial community CO_2 exchange with the atmosphere. Each of these two major categories of research is addressed below.

A Model of Carbon Flow in Terrestrial Microbial Communities

Sources and fate of organic carbon nutrients in soil microbial communities is an important consideration in accounting for the metabolic fate of organic compounds yielding respiratory CO_2. Boddy, et al. (2007, 2008) modeled the dynamics of carbon compounds of varying complexity from sources (e.g., plant soluble and insoluble compounds) through the microbial pools (Figure 2). The fate of amino acids and glucose labeled with ^{14}C was examined in experimental studies using a biphasic model described by a two-process, first order decay equation, fundamentally a refined expression of eq. 1 based on physiological ecology perspectives:

$$S = [a_1 \cdot \exp(-k_1 t)] + [a_2 \cdot \exp(-k_2 t)] \qquad (2)$$

Table 1. Carbon content, and potential respiratory CO_2 efflux of bacteria and protists from Toolik, Alaska during June and August 2007a.

	June			August		
	µg C/g X 10^2	mg C/m^3 /m^2/h	µmol CO_2	µg C/g X 10^2	mg C/m^3 /m^2/h	µmol CO_2
Bacteria	84	40	300	437	200	1,500
Microflagellates	13	6.5	213	53	24	850
Naked Amoebae	3.4	1.6	5.5	0.6	0.3	0.9
Testate Amoebae	5.8	2.8	1.4	41	20	25

[a]The abundances, expressed per g soil mean dry weight for June and august, are respectively: bacteria (4 x10^9 and 2 x10^{10}), microflagellates (1 x 10^7 and 4 x 10^7), naked amoebae (2 x10^4 and 3 x10^4), and testate amoebae (1 x10^3 and 6 x10^3). The larger mean sizes of the testate amoebae contributed to their larger data entries. Data are calculated for a moss-rich soil 2 cm thick and respiration data are normalized for a temperature of 20 °C. Area estimates are based on dry moss density (g/cm^2). Carbon mg/m^3 of soil is rounded to the nearest tenth. From Anderson (2008) with permission.

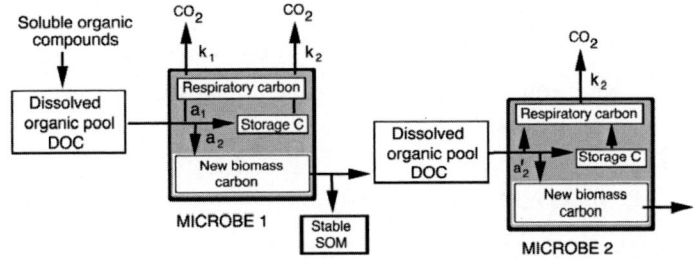

Figure 2. Fate of organic soluble compounds as dissolved organic carbon (DOC) in the soil during uptake and passage through microbes of the belowground microbial community. Initial uptake by microbe 1 in pool a_1 can be relatively rapid serving as substrate for respiratory loss as CO_2 with rate k_1 or by assimilation into the microbial biomass as structural and storage carbon compounds (a_2), subsequently respired at a slower rate k_2. Some of the carbon can be lost as soluble organic matter (SOM) such as waste products, or taken up by predation or assimilation by others (microbe 2) into additional pools a_2' leading to another round of respiratory loss at the lower rate k_2 characteristic of organic carbon passing through the microbial metabolic pool. Based on Boddy, et al. (2007).

where S is the ^{14}C-label remaining in the soil, k_1 is the exponential coefficient describing the primary mineralization by the microbial biomass, k_2 is the exponential coefficient describing the secondary, slower phase of mineralization, a_1 and a_2 describe the proportion of ^{14}C associated with pools with exponential coefficients k_1 and k_2, and t is time. The first rapid phase of $^{14}CO_2$ production is attributable to the immediate use of the substrate in catabolic processes; i.e., respiration. The half-life ($t_{1/2}$) of the pool designated by a_1 can be calculated as $t_{1/2} = \ln(2)/k_1$. The slower second phase (k_2) of $^{14}CO_2$ production was attributed to the subsequent turnover of ^{14}C that had been assimilated and immobilized within the soil microbial community. Various models have been suggested for the fate of C substrates after entering the soil, with varied theories for the connectivity of pool a_2 to a_1 (e.g., Boddy et al., 2007). Given the uncertainty about the connectivity between pools a_1 and a_2, it is not possible, currently, to estimate the half-life for pool a_2. Estimates at 20 °C of the exponential coefficient (k_1) for glucose and amino acids were 0.93 ± 0.07 and 0.64 ± 0.06 h^{-1}, respectively with relatively short half-life ($t_{1/2}$) for each of 1.07 and 1.63 h, respectively. In contrast, the second phase, representing carbon immobilized in the microbial biomass prior to mineralization was much slower (mean $k_2 = 1.30 \times 10^{-3} \pm 0.49 \times 10^{-4}$ h^{-1}). The values of a_1 and a_2 were significantly affected by temperature, field site and type of carbon source. For example, at 4 °C, $a_1 = 9.8$, at 10 °C, $a_1 = 10.7$, while at 20 °C, $a_1 = 12.8$. In general, the model indicates that initial release of soluble organic compounds from vegetation and other sources passes relatively rapidly into the microbial pool and then assumes a substantially longer residence time before transformation and final mineralization takes place. In the process of initial uptake, however, some of the lost carbon is respired as CO_2 or incorporated into biomass; while a portion of it enters the soil soluble organic matter (SOM) pool. The residual carbon represented by the second phase of the model is transferred laterally as DOC to other microbes and/or becomes sequestered by predation in the increasingly higher trophic levels of the food web until mineralization is complete.

Modeling Climate and Vegetation Effects on Soil Respiratory CO_2 Flux

Global and high-latitude regional studies. Models of global respiratory CO_2 flux have been proposed based on regression analyses (e.g., Raich and Schlesinger, 1992; Peng and Apps, 2000) relating respiratory CO_2 to a variety of climatic variables. Peng and Apps (2000) used stepwise regression to explore the relationship between six climatic variables and global soil-CO_2 flux. Among the six independent variables, only temperature, precipitation and actual evapotranspiration correlated significantly with soil-CO_2 flux. The remainders were found to be poorer predictors. Similar regression analyses have been applied to high latitude data to quantitatively model the relationship of soil respiratory CO_2 flux to climate and environmental variables at tundra sites. Illeris and Jonasson (1999) examined the seasonal variations in ecosystem CO_2 emission in a dry subarctic heath in northern Scandinavia. Their regression model showed that respiration rates in this relatively dry tundra were strongly controlled primarily by soil moisture conditions. However, they predicted that CO_2 efflux is likely to increase with increased moisture levels and decrease with drying, which is contrary to expected responses in wet tundra, where excessive water may create anaerobic environments. Ostendorf (1996) modeled spatial and temporal patterns of CO_2 efflux from arctic tundra soils using three, linked simulation models at a 2.2-km^2 catchment site using a suite of geographic-based, meteorological and vegetation canopy models coupled with an empirical regression model incorporating the effects of soil temperature and depth to the water table. Estimated efflux in riparian zones was 60 g C m^{-2} compared to 119 g C m^{-2} in the hillslopes, indicating large spatial differences. This model predicts that an increase of air temperature and solar radiation, or a decrease of precipitation, will increase soil respiration at these particular sites. Within the limitations of the model cited by the author, his results indicate a tight connection between water and carbon cycles at the catchment scale. Keeping all other conditions constant, a seasonal increase of transpiration rates by 10% increases soil respiration by 5% or 4.6 g C m^{-2}.

Further regional level research, especially related to hydrology, was done by Oberbauer, et al. (1992) who examined the contribution of temperature and water table depth in predicting wet tundra CO_2 efflux. In addition to temperature, the water table depth, as noted above, has been shown to be a significant variable in predicting respiratory CO_2 flux to the atmosphere. Some details of the rationale and equations are presented to provide a flavor of their approach. They assumed excessive water that saturates the soil produces anaerobic conditions (among other effects) that may inhibit respiratory CO_2 production. CO_2 efflux data were fitted to a model (eq. 3) incorporating the Arrhenius function for temperature and an asymptotic function (eq. 4) for depth to the water table. An asymptotic function was chosen based on an inspection of the data, which suggested that there was little effect at depths to the water table greater than 10 cm.

$$R_s = C \cdot \exp(-E/RT_k) \cdot \exp(S_{wt}) \qquad (3)$$

where C is a constant providing units of µmol m^{-2} s^{-1}, R_s is rate of CO_2 efflux (µmol m^{-2} s^{-1}), R is the gas constant (8.31 J mol^{-1} °K^{-1}), T_k is the soil temperature (°K) at 1 cm depth, E is the apparent activation energy (J mol^{-1}), and S_{wt} is an empirical function representing the effect of soil water table depth. C is estimated from their data to be approximately 7.8 x 10^{-6}.

$$S_{wt} = A \cdot W_t / (W_t + B) \qquad (4)$$

where W_t is depth to water table below soil surface (cm) and A and B are regression coefficients. Parameters E, A, and B were estimated using nonlinear least-squares regression. For a *Carex* site and an *Eriophorum* site, the respective values were E = 29,144 and 71,178, A = 13.319 and 30.359, and B = 0.321 and 0.044. Non-linear regression modeling of CO_2 has been recommended for assessing efflux in closed chambers where steady state conditions may not be achieved during gas monitoring (e.g., Kutzbach, et al., 2007). The above non-linear model, derived from empirical data collected using sealed plastic tube chambers inserted into the soil, correlated well with actual measured values of R. Values converted to g C m^{-2} d^{-1} ranged from approximately 5 (June) to 11 (July) at the *Carex* site, and from 1 (June) to 9 (July) at the *Eriophorum* site. Additional research by Vourlitis, et al. (2000a) confirmed the significant role of water table depth. They reported that partial regression analysis revealed that water-table depth explained relatively more of the variance in respiratory CO_2 release (45%) than temperature (11%).

CO_2 diffusion and efflux models. To further examine the dynamics of soil porosity and water content on CO_2 diffusion within soils and its flux to the atmosphere, Elberling, et al, (2004) applied an appropriate steady-state diffusion model using laboratory measured production of CO_2 as an input. The model is fundamentally an application **of Fick's second law** of diffusion with an effective diffusion coefficient (D_c) estimated from the variables of air-filled porosity, soil porosity and temperature, among others. They assumed that due to the constraints of soil structure, and the substantial respiratory CO_2 production by roots and microbiota, the gas-filled pores in soils typically have CO_2 concentrations 10 to 100 times higher than the atmosphere, thus creating a strong diffusion gradient. The transport of the near-surface CO_2 is considered largely as a result of diffusion driven by concentration gradients and limited by the decrease in continuous air-filled pores with increasing water content. They applied the model to four high arctic sites varying in moisture content. Simulated CO_2 profiles and CO_2 effluxes (up to 3 μmol CO_2 m^{-2} s^{-1}) agreed with field observations. Moreover, their data indicated the importance of both vegetation- and depth-specific CO_2 production and CO_2 diffusion to fully explain the spatial variation in near-surface soil CO_2 gas dynamics. Furthermore, application of the model confirms that molecular diffusion dominates gas transport, at least in the studied soils.

Complex models, scaling-up regional models. Other models, incorporating primary production and soil respiratory CO_2 efflux, have been devised to permit scaling up local plot measurements to larger geographic regimes (e.g., Vourlitis, et al., 2000b). This research, across a range of arctic tundra sites, combined data from chamber measurements of CO_2 exchange, meteorology, hydrology, and surface reflectance combined with fundamental physiological models of biotic responses to quantify the diurnal and seasonal dynamics of whole-ecosystem respiration (R), gross primary production (GPP), and net CO_2 exchange (F) of wet- and moist-sedge tundra ecosystems of arctic Alaska. The diurnal fluctuations in R were expressed as exponential functions of air temperature. Diurnal fluctuations in GPP were described as hyperbolic functions of diurnal photosynthetic photon flux density (PPFD). Daily integrated rates of R were expressed as an exponential function of average daily water table depth and temperature; whereas, daily fluctuations in GPP were described as a hyperbolic function of average daily PPFD and a sigmoidal function of the normalized difference vegetation index (NDVI) calculated from satellite imagery. Their models described, on average, 75-97% of the variance in diurnal R and GPP, and 78-95% of the variance in total daily R and GPP. Moreover, their data, scaled up nicely to hectare-size areas, and estimated seasonally integrated CO_2 exchange within 20% of the observed value based on eddy covariance measurements.

Temperature-Dependent Estimations of Microbial Respiratory CO_2 Flux

Although some intensive research has been done to estimate composite microbial contributions to total soil respiratory CO_2 flux, little is known about the contribution made in the composite by individual major taxonomic groups of microbes, especially at high latitudes. This is due mainly to prior lack of knowledge about the abundances and carbon content of the individual taxa. Recently, Anderson (2008), using regression equations relating cell carbon content and respiratory CO_2 efflux to cell volume, enumerated bacterial and major eukaryotic microbial groups in tundra surface soil from Toolik, Alaska and estimated the amount of CO_2 released by each taxon and their total amount for two sampling dates in June and August of 2007 (assuming a standard reference temperature of 20 °C). Bacteria, heterotrophic small flagellates, naked amoebae and testate amoebae were enumerated and sized microscopically for each sample, and the data were used to compute carbon content and respiratory release (e.g., Table 1). In addition, the likely variation in total respiration, with different proportions of active vs. dormant bacteria (putatively related especially to moisture content), was calculated and plotted as a family of linear successive plotted lines related to bioactive soil depth (Figure 3a,b). The latter, presumably, is largely related to the depth of thaw. The values for percent active bacteria is plotted only for a range of 50 – 100% to simplify the graph. As an initial approximation, each successive plotted point (function of depth) is a linear sum of the values beneath it. For example, for a 50% active bacterial population at 10 cm depth (Figure 3b), the estimated CO_2 flux is approximately 8,000 mol km^{-2} hour^{-1}, which is twice the amount predicted for a five cm depth. Based on the data in Table 1, the following equations were derived to produce the plots in Figure 3, beginning with the general expression of the equation.

$$R = [r_p + (r_b \cdot P_b)] \cdot D_b \qquad (5)$$

where R is the respiration rate (mol km^{-2} h^{-1}), r_p is the protistan respiratory contribution, r_b is the bacterial contribution, P_b is the proportion of active bacteria, and D_b is the bioactive depth in cm. Solutions for a given D_b value and a range of P_b values generates a set of Y-axis coordinates for a line plotted at that bioactive depth (Figure 3).

$$R_J = [110 + (150 \cdot P_b)] \cdot D_b \qquad (6)$$

$$R_A = [438 + (750 \cdot P_b)] \cdot D_b \qquad (7)$$

where R_J and R_A are respectively the respiration rates for the June and August samples, respectively.

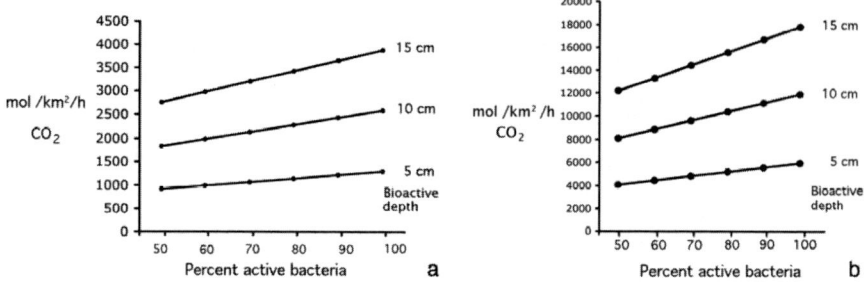

Figures 3a, b. First approximations to linear models of temperature-dependent, microbial respiratory CO_2 flux to the atmosphere based on data derived from soil samples from Toolik, Alaska in June and August (2007) and compiled from data in Table 1. Figure 3a, estimated flux (ordinate) for a June sample in relation to percent active bacteria (abscissa) within a limited range of 50 to 100% and at varying bioactive (thaw) depths (cm) presented as a family of linear plots. Figure 3b is a similar graph for the August data based on the larger population sizes during that warmer portion of the seasonal cycle. Data are normalized for a temperature of 20 °C. From Anderson (2008) with permission.

This model, with a family of successive higher-level, linear plots, must be considered as only a first approximation. Current evidence, as summarized above in this chapter, indicates that abundances of microbial taxa are not uniformly distributed in the soil horizons and overall decrease with depth. Hence, assuming we obtain more exact information on depth distribution of microbiota, a depth-limited, decreasing, nonlinear expression may need to be applied for D_b and perhaps partitioned as different values for protistan and bacterial terms to better approximate their individual abundances with depth. The data in this draft of the model are also limited by using a standardized temperature of 20 °C, where we have good estimates of respiratory activity. If sufficiently accurate estimates of the Arrhenius equation coefficients can be determined for various microbial taxa, a more exact determination of temperature-related metabolic activity and respiratory CO_2 release may be possible. However, the model provides one of the first estimates of carbon content and likely respiratory CO_2 release by Protista as well as bacterial microbial taxa in a tundra soil community.

Overall, these data indicate that with increasing bioactive depth and sufficient moisture to support active bacterial and protistan metabolism, substantial increases in soil respiratory efflux may occur with increased global warming. This model, limited to bacterial and major terrestrial protists, should also be augmented by better estimates of fungal contributions to the carbon budget and respiratory CO_2 efflux. Given the substantial fungal biomass in tundra soils at some locations, as documented in preceding sections of this chapter, a more complete estimation of microbial community contribution to the carbon cycle will depend on better estimates of fungal biomass and metabolic activity. One of the limitations in estimating microorganism respiration is the large variability in response to variations in the quality and kind of metabolized substrate. The algorithms are presently rather unrefined. The estimates of carbon content may be more robust and provide evidence of the relative carbon content among the four major groups listed in Table 1. The current estimates of respiration (Anderson, 2008) for bacterial and protistan contributions (agreeing favorably with laboratory measured total CO_2 release from sieved tundra soil) provide a step forward toward a more complete account of total microbial respiratory CO_2 release in high latitude ecosystems. In general, the comparative data for the major groups of microbes studied here (Table 1) are probably sufficiently robust to provide some of the first evidence of the importance of including the contributions of eukaryotic microbes when computing the carbon budgets of microbial communities, especially at higher latitudes. Models of this kind, or other refinements of them, may be useful components of more elaborate and comprehensive models that integrate data across all major biota in the high latitude soil environment (including microfaunal, microbial, and root contributions).

CONCLUSION

There is clear evidence that substantial carbon reserves are stored in vast geographic areas of the tundra within permafrost soil and constituent organic matter that has accumulated over millennia by the deposition of organic phytomass. With predictions of increased global warming, there is a growing concern that the current net respiratory CO_2 sink in many of these high latitude biomes may become a net CO_2 source, especially if belowground metabolic respiration by plant roots, microfauna and particularly microbiota increases as the permafrost melts and the soils warm. Moreover, there is current evidence that belowground respiration continues at an appreciable degree even during the snow-cover seasons, thus potentially summing with increased snow-free season efflux to produce a substantial increase in atmospheric CO_2 concentrations. By contrast, if the soil is water-saturated due to local hydrological conditions and becomes anaerobic, respiratory CO_2 may be depressed. However, the potential for release of methane gas by methanogens could increase, thus contributing to an elevated greenhouse effect. This is a topic deserving much more systematic inquiry.

Our ability to accurately predict future scenarios is limited by several factors, including: sufficient detailed information about regional climatic changes that are likely to occur using existing climate models, uncertainties about precipitation patterns and the hydrology of tundra ecosystems, and less than sufficient data about the relative contributions of the various belowground biota to the total respiratory CO_2 flux and how it may change with changing climatic conditions. There is particular interest in better estimation of water table depth that, if sufficiently shallow, may create anaerobic conditions limiting respiratory CO_2 release. On the other hand, dry tundra sites may experience increased respiratory CO_2 flux if more favorable precipitation patterns emerge.

Overall, there are some promising current and emerging models relating respiratory CO_2 flux to vegetation, regional geographic data, climate, hydrology, and coupling of belowground communities to aboveground climatic and ecological processes. However, our limited knowledge of the belowground biological communities, and their complex relationships to one another and to belowground phytomass, is a major barrier to a more complete prediction of possible effects of future global warming on tundra ecosystems. Much additional systems-level research is needed in this significant area of high-latitude science.

ACKNOWLEDGMENT

My sincere thanks to Prof. Kevin Griffin for reviewing this contribution and making helpful suggestions. This is Lamont-Doherty Contribution Number 7263.

REFERENCES

Aber, J. D., Melillo, J. M. & McClaugherty, C. A. (1990). Predicting long-term patterns of mass loss, nitrogen dynamics, and soil organic matter formation from initial fine litter chemistry in temperate forest ecosystems. *Canadian Journal of Botany, 68,* 2201-2208.

Ananyeva, N. D., Susyan, E. A., Chernova, O. V., Chernov, I. Yu & Makarova, O. L. (2006). The ratio of fungi and bacteria in the biomass of different types of soil determined by selective inhibition. *Microbiology, 75,* 702-707.

Anderson, O. R. & Griffin, K. (2001). Abundances of protozoa in soil of laboratory-grown wheat plants cultivated under low and high atmospheric CO_2 concentrations. *Protistology, 2,* 76-84.

Anderson, O. R. (2006a). A method for estimating cell volume of amoebae based on measurements of cell length of motile forms: physiological and ecological applications. *Journal of Eukaryotic Microbiology, 53,* 185-187.

Anderson, O. R. (2006b). The Density and Diversity of Gymnamoebae Associated with Terrestrial Moss Communities (Bryophyta: Bryopsida) in a Northeastern U.S. Forest. *Journal of Eukaryotic Microbiology, 53,* 275-279.

Anderson, O. R. (2007). A seasonal study of the carbon content of planktonic naked amoebae in the Hudson Estuary and in a productive freshwater pond with comparative data for ciliates. *Journal of Eukaryotic Microbiology, 54,* 388-391.

Anderson, O. R. (2008). The Role of amoeboid protists and the microbial community in moss-rich terrestrial ecosystems: Biogeochemical implications for the carbon budget and carbon cycle, especially at higher latitudes. *Journal of Eukaryotic. Microbiology, 55,* 145-150.

Beyens, L., Ledeganck, P., Graae, B. J. & Nijs, I. (2009). Are soil biota buffered against climatic extremes? An experimental test on testate amoebae in arctic tundra (Qeqertarsuag, West Greenland). *Polar Biology, 32,* 453-462.

Billings, W. D., Luken, J. O., Mortenson, D. A. & Peterson, K, M, (1983). Increasing atmospheric carbon dioxide: possible effects on arctic tundra. *Oecologia, 58,* 286-289.

Billings, W. D., Luken, J. O., Mortenson, D. A. & Peterson, K. M. (1982). Arctic tundra: a source or sink for atmospheric carbon dioxide in a changing environment. *Oecologia, 53,* 7-11.

Bjork, R. G., Bjorkman, M. P., Andersson; M. X. & Klemedtsson, L. (2008). Temporal variation in soil microbial communities in Alpine tundra. *Soil Biology and Biochemistry, 40,* 266-268.

Bockheim, J. G., Everett, L. R., Hinkel, K. M., Nelson, F. E., & Brown, J. (1999). Soil organic carbon storage and distribution in arctic tundra, Barrow, Alaska. *Soil Science Society of America Journal, 63,* 934-940.

Boddy, E., Hill, P. W., Farrar, J. & Jones, D. L. (2007). Fast turnover of low molecular weight components of the dissolved organic carbon pool of temperate grassland field soils. *Soil Biology and Biochemistry, 39*, 827-835.

Boddy, E., Roberts, P., Hill, P. W., Farrar, J. & Jones, D. L. (2008). Turnover of low molecular weight dissolved organic C (DOC) and microbial C exhibits different temperature sensitivities in Arctic tundra soils. *Soil Biology and Biochemistry, 40*, 1557-1566.

Broecker, W. S. (1975). Climatic change: are we on the brink of a pronounced global warming? *Science, 189*, 460-463.

Buckeridge, K. M. & Grogan, P. (2008). Deepened snow alters soil microbial nutrient limitations in arctic birch hummock tundra. *Applied Soil Ecology, 39*, 210-222.

Calef, M. P., McGuire, A. D., Epstein, H. E., Rupp, T. S. & Shugart, H. H. (2005). Analysis of vegetation distribution in Interior Alaska and sensitivity to climate change using a logistic regression approach. *Journal of Biogeography, 32*, 863-878.

Callaghan, T. V., Bjorn, L. O., Chapin, T., Christensen, T. R., Huntley, B., Ims, R. A., Johansson, M., Jolly, D., Jonasson, S., Matveyeva, N., Panikov, N., Oechel, W., Shaver, G. & Henttonen, H. (2004). Effects on the structure of arctic ecosystems in the short- and long-term perspectives. *Ambio, 33*, 436-447.

Chapin, F. S., III; Shaver, G. R., Giblin, A. E., Nadelhoffer, K. J. & Laundre, J. A. (1995). Response of arctic tundra to experimental and observed changes in climate. *Ecology, 76*, 694-711.

Coulson, S. J., Hodkinson, I. D., Webb, N. R., Block, W., Bale, J. S., Strathdee, A. T., Worland, M. R. & Wooley, C. (1996). Effects of experimental temperature elevation on high-arctic soil microarthropod populations. *Polar Biology, 16*, 147-153.

Couteaux, M.-M. & Bolger, T. (2000). Interactions between atmospheric CO_2 enrichment and soil fauna. *Plant and Soil, 224*, 123-134.

Drigo, B., Kowalchuk, G. A. & van Veen, J. A. (2008). Climate change goes underground: effects of elevated atmospheric CO_2 on microbial community structure and activities in the rhizosphere. *Biology and Fertility of Soils, 44*, 667-679.

Ekelund, F., Ronn, R. & Christensen, S. (2001). Distribution with depth of protozoa, bacteria and fungi in soil profiles from three Danish forest sites. *Soil Biology and Biochemistry, 33*, 475-481.

Elberling, B., Jakobsen, B. H., Berg, P., Sondergaard, J. & Sigsgaard, C. (2004). Influence of vegetation, temperature, and water content on soil carbon distribution and mineralization in four high Arctic soils. *Arctic Antarctic and Alpine Research, 36*, 528-538.

Eugster, W., McFadden, J. P. & Chapin, F. S. III. (2005). Differences in surface roughness, energy, and CO_2 fluxes in two moist tundra vegetation types, Kuparuk watershed, Alaska, USA. *Arctic Antarctic and Alpine Research, 37*, 61-67.

Grogan, P. & Jonasson, S. (2005). Temperature and substrate controls on intra-annual variation in ecosystem respiration in two subarctic vegetation types. *Global Change Biology, 11*, 465-475.

Gundelwein, A., Mueller-Lupp, T., Sommerkorn, M., Haupt, E. T. K., Pfeiffer, E.-M. & Wiechmann, H. (2007). Carbon in tundra soils in the Lake Labaz region of arctic Siberia. *European Journal of Soil Science, 58*, 1164-1174.

Harding, R., Kuhry, P., Christensen, T. R., Sykes, M. T., Dankers, R. & van der Linden, S. (2002). Climate feedbacks at the tundra-taiga interface. *Ambio Special Report, 12*, 47-55.

Hobbie, J. E. & Rublee, P. A. (1999). Controls on microbial food webs in oligotrophic arctic lakes. *Ergebnisse der Limnologie, 54*, 61-76.

Hobbie, S. E. & Chapin, F. S. III. (1998). The response of tundra plant biomass, aboveground production, nitrogen, and CO_2 flux to experimental warming. *Ecology, 79*, 1526-1544.

Illeris, L. & Jonasson, S. (1999). Soil and plant CO_2 emission in response to variations in soil moisture and temperature and to amendment with nitrogen, phosphorus, and carbon in northern Scandinavia. *Arctic Antarctic and Alpine Research, 31*, 264-271.

Jones, M. H., Fahnestock, J. T., Walker, D. A., Walker, M. D. & Welker, J. M. (1998). Carbon dioxide fluxes in moist and dry Arctic tundra during the snow-free season: Responses to increases in summer temperature and winter snow accumulation. *Arctic and Alpine Research, 30*, 373-380.

Kaiser, C., Meyer, H., Biasi, C., Rusalimova, O., Barsukov, P. & Richter, A. (2005). Storage and mineralization of carbon and nitrogen in soils of a frost-boil tundra ecosystem in Siberia. *Applied Soil Ecology, 29*, 173-183.

Kolchugina, T. & Vinson, T. S. (1993). Carbon balance of the continuous permafrost zone of Russia. *Climate Research, 3*, 13-21.

Kotsyurbenko, O. R., Nozhevnikova, A. N., Soloviova, T. I. & Zavarzin, G. A. (1996). Methanogenesis at low temperature by microflora of tundra wetland soil. *Antonie van Leeuwenhoek, 69*, 75-86.

Kutzbach, L. (2006). The exchange of energy, water and carbon dioxide between wet arctic tundra and the atmosphere at the Lena River Delta, Northern Siberia. *Berichte zur Polar- und Meeresforschung, 541*, 1-141, III, IV.

Kutzbach, L., Schneider, J., Sachs, T., Giebels, M., Nykanen, H., Shurpali, N. J., Martikainen, P. J., Alm, J. & Wilmking, M. (2007). CO_2 flux determination by closed-chamber methods can be seriously biased by inappropriate application of linear regression. *Biogeosciences, 4*, 1005-1025.

Lafleur, P. M., Griffis, T. J. & Rouse, W. R. (2001). Interannual variability in net ecosystem CO_2 exchange at the arctic treeline. *Arctic Antarctic and Alpine Research, 33*, 149-157.

Lichter, J., Barron, S. H., Bevacqua, C. E., Finzli, A. C., Irving, K. F., Stemmler, E. A. & Schlesinger, W. H. (2005). Soil carbon sequestration and turnover in a pine forest after six years of atmospheric CO_2 enrichment. *Ecology, 86*, 1835-1847.

Loya, W. M. & Grogan, P. (2004). Global change: Carbon conundrum on the tundra. *Nature, 431*, 406-408.

Luken, J. O. & Billings, W. D. (1985). The influence of microtopographic heterogeneity on carbon dioxide efflux from a subarctic bog. *Holarctic Ecology, 8*, 306-312.

Mano, M., Harazono, Y., Miyata, A. Z., Rommel C. & Oechel, W. C. (2003). Net CO_2 budget and seasonal variation of CO_2 fluxes at a wet sedge tundra ecosystem at Barrow, Alaska during the 2000 growing season. *Journal of Agricultural Meteorology, 59*, 141-154.

Masek, J. G. (2001). Stability of boreal forest stands during recent climate change: Evidence from Landsat satellite imagery. *Journal of Biogeography, 28*, 967-976.

McFadden, J. P., Eugster, W. & Chapin, F. S. III. (2003). A regional study of the controls on water vapor and CO_2 exchange in arctic tundra. *Ecology, 84*, 2762-2776.

McGuire, A. D., Wirth, C., Apps, M., Beringer, J., Clein, J., Epstein, H., Kicklighter, D. W., Bhatti, J., Chapin, F. S. III; de Groot, B., Efremov, D., Eugster, W., Fukuda, M., Gower, T., Hinzman, L., Huntley, B., Jia, G. J., Kasischke, E., Melillo, J., Romanovsky, V., Shvidenko, A., Vaganov, E. & Walker, D. (2002). Environmental variation, vegetation distribution, carbon dynamics and water/energy exchange at high latitudes. *Journal of Vegetation Science, 13*, 301-314.

McMichael, C. E., Hope, A. S., Stow, D. A., Fleming, J. B., Vourlitis, G. & Oechel, W. (1999). Estimating CO_2 exchange at two sites in Arctic tundra ecosystems during the growing season using a spectral vegetation index. *International Journal of Remote Sensing, 20*, 683-698.

Mertens, S., Nijs, I., Heuer, M., Kockelbergh, F., Beyens, L., Van Kerckvoorde, A. & Impens, I. (2001). Influence of high temperature on end-of-season tundra CO_2 exchange. *Ecosystems, 4*, 226-236.

Meyer, H., Kaiser, C., Biasi, C., Haemmerle, R., Rusalimova, O., Lashchinsky, N., Baranyi, C., Daims, H., Barsukov, P. & Richter, A. (2006). Soil carbon and nitrogen dynamics along a latitudinal transect in Western Siberia, Russia. *Biogeochemistry, 81*, 239-252.

Neufeld, J. D. & Mohn, W. W. (2005). Unexpectedly high bacterial diversity in arctic tundra relative to boreal forest soils, revealed by serial analysis of ribosomal sequence tags. *Applied and Environmental Microbiology, 71*, 5710-5718.

Nobrega, S. & Grogan, P. (2008). Landscape and ecosystem-level controls on net carbon dioxide exchange along a natural moisture gradient in Canadian low arctic tundra. *Ecosystems, 11*, 377-396.

Nordin, A., Schmidt, I. K. & Shaver, G. R. (2004). Nitrogen uptake by arctic soil microbes and plants in relation to soil nitrogen supply. *Ecology, 85*, 955-962.

Nozhevnikova, A. N., Rebak, S., Kotsyurbenko, O. R., Parshina, S. N., Holliger, C. & Lettinga, G. (2000). Anaerobic production and degradation of volatile fatty acids in low temperature environments. *Water Science and Technology, 41*, 39-46.

O'Neill, K. P. (2000). Role of bryophyte-dominated ecosystems in the global carbon budget. In: A. J. Shaw, & B. Goffinet (Eds.), *Bryophyte Biology* (344-368). Cambridge, UK: Cambridge University Press.

Oberbauer, S. F., Gillespie, C. T., Cheng, W., Gebauer, R., Sala Serra, A. & Tenhunen, J. D. (1992). Environmental effects on carbon dioxide efflux from riparian tundra in the northern foothills of the Brooks Range, Alaska, USA. *Oecologia, 92*, 568-577.

Oberbauer, S. F., Oechel, W. C. & Riechers, G. H. (1986). Soil respiration of Alaskan USA tundra at elevated atmospheric carbon dioxide concentrations. *Plant and Soil, 96*, 145-148.

Oberbauer, S. F., Tenhunen, J. D. & Reynolds, J. F. (1991). Environmental effects on carbon dioxide efflux from water track and tussock tundra in arctic Alaska USA. *Arctic and Alpine Research, 23*, 162-169.

Oechel, W. C., Cowles, S., Grulke, N., Hastings, S. J., Lawrence, B., Prudhomme, T., Riechers, G., Strain, B., Tissue, D. & Vourlitis, G. (1994). Transient nature of CO_2 fertilization in Arctic tundra. *Nature, 371*, 500-503.

Oechel, W. C., Vourlitis, G. & Hastings S. J. (1997). Cold season CO_2 emission from arctic soils. *Global Biogeochemical Cycles, 11*, 163-172.

Oechel, W. C., Vourlitis, G. L., Hastings, S. J. & Bochkarev, S. A. (1995). Change in arctic CO_2 flux over two decades: Effects of climate change at Barrow, Alaska. *Ecological Applications, 5*, 846-855.

Oechel, W. C., Vourlitis, G. L., Hastings, S. J., Zulueta, R. C., Hinzman, L. & Kane, D. (2000). Acclimation of ecosystem CO_2 exchange in the Alaskan Arctic in response to decadal climate warming. *Nature, 406*, 978-981.

Ostendorf, B. (1996). Modeling the influence of hydrological processes on spatial and temporal patterns of CO_2 soil efflux from an arctic tundra catchment. *Arctic and Alpine Research, 28*, 318-327.

Peng, C-h. & Apps, M. J. (2000). Simulating global soil-CO_2 flux and its response to climate change. *Journal of Environmental Sciences, 12*, 257-265.

Ping, C. L., Michaelson, G. J. & Kimble, J. M. (1997). Carbon storage along a latitudinal transect in Alaska. *Nutrient Cycling in Agroecosystems, 49*, 235-242.

Raich, J. W. & Schlesinger, W. H. (1992). The global carbon dioxide flux in soil respiration and its relationship to vegetation and climate. *Tellus Series B Chemical & Physical Meteorology, 44*, 81-99.

Rien, A. (2009). Nitrogen supply effects on leaf dynamics and nutrient input into the soil of plant species in sub-arctic tundra ecosystems. *Polar Biology, 32*, 207-214.

Rustad, L. E., Campbell, J. L., Marion, G. M., Norby, R. J., Mitchell, M. J., Hartley, A. E., Cornelissen, J. H. C., Gurevitch, J. & GCTE-NEWS (2001). A meta-analysis of the response of soil respiration, net nitrogen mineralization, and aboveground plant growth to experimental ecosystem warming. *Oecologia, 126*, 543-562.

Sawstrom, C., Mumford, P., Marshall, W., Hodson, A. & Laybourn-Parry, J. (2002). The microbial communities and primary productivity of cryoconite holes in an Arctic glacier (Svalbard 79 degree N). *Polar Biology, 25*, 591-596.

Schimel, J. P. & Chapin, F. S., III. (1996). Tundra plant uptake of amino acid and NH4+ nitrogen in situ: Plants compete will for amino acid N. *Ecology, 77*, 2142-2147.

Schimel, J.P. & Mikan, C. (2005). Changing microbial substrate use in Arctic tundra soils through a freeze–thaw cycle. *Soil Biology and Biochemistry, 37*, 1411-1418.

Schmidt, I. K., Jonasson, S., Shaver, G. R., Michelsen, A. & Nordin, A. (2002). Mineralization and distribution of nutrients in plants and microbes in four arctic ecosystems: Responses to warming. *Plant and Soil, 242*, 93-106.

Schmidt, K., Lipson, D. A., Ley, R. E., Fisk, M. C. & West, A. E. (2004). Impacts of chronic nitrogen additions vary seasonally and by microbial functional group in tundra soils. *Biogeochemistry*, 69, 1-17.

Schmidt, N. & Boelter, M. (2002). Fungal and bacterial biomass in tundra soils along an arctic transect from Taimyr Peninsula, central Siberia. *Polar Biology*, 25, 871-877.

Schröter, D., Brussaard, L., De Deyn, G., Poveda, K., Brown, V. K., Berg, M. P., Wardle, D. A., Moore, J. & Wall, D. H. (2004). Trophic interactions in a changing world: modelling aboveground–belowground interactions. *Basic and Applied Ecology*, 5, 515-528.

Shaver, G. R. & Jonasson, S. (1999). Response of Arctic ecosystems to climate change: Results of long-term field experiments in Sweden and Alaska. *Polar Research*, 18, 245-252.

Shaver, G. R., Giblin, A. E., Nadelhoffer, K. J., Thieler, K. K., Downs, M. R., Laundre, J. A. & Rastetter, E. B. (2006). Carbon turnover in Alaskan tundra soils: effects of organic matter quality, temperature, moisture and fertilizer. *Journal of Ecology*, 94, 740-753.

Sjogersten, S., van der Wal, R. & Woodin, S. J. (2008). Habitat type determines herbivory controls over CO_2 fluxes in a warmer arctic. *Ecology*, 89, 2103-2116.

Skre, O., Baxter, R., Crawford, R. M. M., Callaghan, T. V. & Fedorkov, A. (2002). How will the tundra-taiga interface respond to climate change? *Ambio Special Report*, 12, 37-46.

Sommerkorn, M. (1998) Patterns and controls of CO_2 fluxes in wet tundra types of the Taimyr Peninsula, Siberia – the contribution of soils and mosses. *Berichte zur Polarforschung*, 298, 1-219.

Sorensen, P. L., Michelsen, A. & Jonasson, S. (2008). Ecosystem partitioning of N15-glycine after long-term climate and nutrient manipulations, plant clipping and addition of labile carbon in a subarctic heath tundra. *Soil Biology and Biochemistry*, 40, 2344-2350.

Stapleton, L. M., Crout, N. M. J., Sawstrom, C., Marshall, W. A., Poulton, P. R., Tye, A. M. & Laybourn-Parry, J. (2005). Microbial carbon dynamics in nitrogen amended Arctic tundra soil: Measurement and model testing. *Soil Biology and Biochemistry*, 37, 2088-2098.

Tate, R. L. III (1995). *Soil Microbiology* (247-249). New York: John Wiley and Sons.

Trofymow, J. A., Preston, C. M. & Prescott, C. E. (1995). Litter quality and its potential effect on decay rates of materials from Canadian forests. *Water, Air and Soil Pollution*, 82, 215-226.

Vourlitis, G. L., Harazono, Y., Oechel, W. C., Yoshimoto, M. & Mano, M. (2000a). Spatial and temporal variations in hectare-scale net CO_2 flux, respiration and gross primary production of Arctic tundra ecosystems. *Functional Ecology*, *14*, 203-214.

Vourlitis, G. L., Oechel, W. C., Hope, A., Stow, D., Boynton, B., Verfaillie, J., Jr., Zulueta, R. & Hastings, S. J. (2000b). Physiological models for scaling plot measurements of CO_2 flux across an arctic tundra landscape. *Ecological Applications*, *10*, 60-72.

Walker, D. A. & Walker, M. D. (1996). Terrain and vegetation of the Imnavait Creek watershed. In J. F. Reynolds, & J. D. Tenhunen (Eds.), Landscape Function and Disturbance in Arctic Tundra (73-108). Berlin: Springer.

Walker, M. D., Wahren, C. H., Hollister, R. D., Henry, G. H. R., Ahlquist, L. E., Alatalo, J. M., Bret-Harte, M. S., Calef, M. P., Callaghan, T. V., Carroll, A. B., Epstein, H. E., Jonsdottir, I. S., Klein, J. A., Magnusson, B., Molau, U., Oberbauer, S. F., Rewa, S. P., Robinson, C. H., Shaver, G. R., Suding, K. N., Thompson, C. C., Tolvanen, A., Totland, O., Turner, P. L., Tweedie, C. E., Webber, P. J. & Wookey, P. A. (2006). Plant community responses to experimental warming across the tundra biome. *Proceedings of the National Academy of Sciences of the United States of America*, *103*, 1342-1346.

Wardle, D. A., Bardgett, R. D., Klironomos, J. N., Setala, H., van der Putten, W. H. & Wall, D. H. (2004). Ecological linkages between aboveground and belowground biota. *Science*, *304*, 1629-1633.

Welker, J. M., Brown, K. B. & Fahnestock, J. T. (1999). CO_2 flux in arctic and alpine dry tundra: Comparative field responses under ambient and experimentally warmed conditions. *Arctic Antarctic and Alpine Research*, *31*, 272-277.

Williams, M. & Rastetter, E. B. (1999). Vegetation characteristics and primary productivity along an arctic transect: Implications for scaling-up. *Journal of Ecology*, *87*, 885-898.

Wilson, J. A. & Coxson, D. S. (1999). Carbon flux in a subalpine spruce-fir forest: pulse release from Hylocomium splendens feather-moss mats. *Canadian Journal of Botany*, *77*, 564-569.

Wolters, V., Silver, W. L., Bignell, D. E., Coleman, D. C., Lavelle, P., van der Putten, W. H., de Ruiter, P., Rusek, J., Wall, D. H., Wardle, D. A., Brussaard, L., Dangerfield, J. M., Brown, V. K., Giller, K. E., Hooper, D. U., Sala, O., Tiedje, J. & van Veen, J. A. (2000). Effects of global changes on above- and belowground biodiversity in terrestrial ecosystems: implications for ecosystem functioning. *Bioscience*, *50*, 1089-1098.

Yoshimoto, M., Harazono, Y. & Oechel, W. C. (1997). Effects of micrometeorology on the CO_2 budget in mid-summer over the Arctic Tundra at Prudhoe Bay, Alaska. *Journal of Agricultural Meteorology, 53*, 1-10.

Zak, D. R. & Kling, G. W. (2006). Microbial community composition and function across an arctic tundra landscape. *Ecology, 87*, 1659-1670.

In: Tundras: Vegetation, Wildlife…   ISBN: 978-1-60876-588-1
Editors: B. Gutierrez et al. pp. 81-110   © 2010 Nova Science Publishers, Inc.

Chapter 3

SOIL AND PLANT CHARACTERISTICS IN THE ALPINE TUNDRA (NW ITALY)

Michele Freppaz[1], Gianluca Filippa[1], Angelo Caimi[1], Giorgio Buffa[2] and Ermanno Zanini[1]

[1]Università degli Studi di Torino, DIVAPRA-Chimica Agraria e Pedologia,
Laboratorio Neve e Suoli Alpini, 44,
Via Leonardo da Vinci, 10095 Grugliasco (TO).
[2]Università degli Studi di Torino, Dipartimento di Biologia vegetale,
25, Viale Mattioli, 10125 Torino.

ABSTRACT

The arctic tundra forms a circumpolar band between the Arctic Ocean and the polar ice caps to the north and the coniferous forests to the south. Smaller, but ecologically similar regions found above the tree line on high mountains are called alpine tundra. Here the soils are perennially cold and often snow-covered and the plants show a high degree of specialization in order to survive in such extreme conditions.

In this work, we aimed to evaluate pedogenetic processes, soil nutrient status and plant distribution along an elevation gradient in the alpine tundra in the western Italian Alps.

The study area for this investigation is located in North West Italy, close to the Monte Rosa Massif (4634 m asl). Soil profiles were dug and described at 5 sites along an elevation gradient from 2525 m asl to 2840 m asl, in the upper part of a glacial valley. Genetic horizons were identified and physical properties described following standard methodology. Bulk samples were taken from each major horizon, and smaller known-volume samples were taken for determination of bulk density. In the laboratory soil chemistry and particle size distribution were determined following standard methods. Data on the vegetation structure were collected close to each soil profile, covering a surface of 32 m^2; each sampling site has been further divided into 8 sub-areas of 4 m^2. The abundances of species were recorded as cover percentages.

Soil depth in the five study areas ranges from 10 to more than 50 cm. Soil profiles exhibit a fairly consistent horizonation. All of the profiles have a black to dark brown A horizon, in some cases underlying thin Oi/Oe horizons. Structure is usually weakly developed fine granular, except at lowest elevation, where is moderately developed subangolar blocky. Textures range from sandy to loamy sand. The major pedogenic processes responsible for soil formation in this environment are: a) organic matter accumulation and melanization; b) cryoturbation; c) erosion and deposition of material. Melanization appeared as a sharp contrast in darkness between the surface horizons and deeper parts of the solum. The accumulation of organic matter is encouraged by low temperatures that reduce rates of decomposition. The presence of irregular horizon boundaries within the profiles are evidence of cryoturbation, as these soils are frost affected, mainly in early winter and late spring, when the reduced snow cover may not be sufficient to insulate the soil from the air temperature. Further, soil erosion and deposition may occur under steep slopes, determining a loss and a subsequent redeposition of soil material and thereby playing a role in pedogenetic processes.

A total of 64 plant species were found at the sampling sites. As expected, the harsh alpine conditions influenced both the life form and the chorology of taxa. Hemicryptophytes (91.07%) were dominant, followed by chamaephytes (6.82%), geophytes (1.21%) and therophytes (0.91%). From the phytogeographical point of view, the species were mainly south european mountain (49.44%), artic alpine (29.74%), west alpine (7.81%) and alpine (6.82%) while the ones with wider range, as european or eurosibiric, were less frequent (6.19%).

Soils described in this area are surprisingly well developed. The organic carbon content of the surface horizons is quite high, and comparable with forested soils at lower elevation. From the top to bottom positions the synecology of plant populations depicted some clear gradients connected to changes in the most important landscape variables: decreasing in bryophyte cover, increasing in vascular plant cover, nitrogen content and in temperature.

INTRODUCTION

The arctic tundra forms a circumpolar band between the Arctic Ocean and the polar ice caps to the north and the coniferous forests to the south. Smaller, but ecologically similar regions found above the tree line on high mountains are called alpine tundra. Alpine tundra occurs across a very great range of latitudes and geological settings, and their environments are correspondingly diverse. Here the soils are perennially cold and often snow-covered and the plants show a high degree of specialization in order to survive in such extreme conditions (Körner, 2003). Permafrost and seasonally frozen ground comprise 24% and 60%, respectively, of the Northern Hemisphere's land surface (Zhang et al., 1999). These perennially cold, often snow-covered soils store a quarter to a half of all soil carbon (C) (Post et al., 1982) and interact with substantial quantities of water destined for human use (Williams et al., 2002).

In ecosystems of cold climates the majority of organically bound plant nutrients are incorporated in the soil organic matter pool. In Alpine tundra ecosystems, most organically bound nutrients are fixed in the soil and litter and a low proportion is in the plant biomass. For instance, the vegetation contained <35% of the ecosystem nitrogen pool, while the remaining 65% was fixed in the recalcitrant soil organic matter and in the litter in Swedish Arctic tundra (Wu et al., 2006). The main reason consists in low microbial activities in extreme ecological conditions as a result of low temperature and high soil acidity leading to slow mineralization of organic compounds (Glaser et al., 2000). Therefore, in spite of the large contents of total nitrogen (N), the concentration of available inorganic N is extremely low, and N availability is a major factor regulating primary plant production and plant community composition in ecosystems of cold climates (Nadelhoffer et al., 1992; Bowman et al., 1993; Makarov et al., 2001).

The Alps are an essential element of the landscape of Central Europe. High-mountain ecosystems are, however, very sensitive to changing environmental conditions such as global warming. In this sense, with regard to vascular plant communities and vegetation formations, the displacement of present vegetation belts towards higher altitudinal or latitudinal areas may serve as a theoretical construct to demonstrate that climate change induces shifts in the vegetation zonation, but is unlikely to occur without major internal reorganisation in terms of both composition and relative proportion of species (Walker and Steffen, 1997). Due to complex and highly species-specific feedback mechanisms, the process of shifting vegetation formations implies the rearrangement of affected assemblages. In the long run this may lead to a severe fragmentation of the communities and habitat loss (Dirnböck et al., 2003). In the same way a warmer climate could give rise to a modified microbiology with the consequence that the weathering regime and SOM storage mechanisms would change (Egli et al., 2004; Edwards et al., 2007; Egli et al., 2009). The soil could shift from being a C source to a sink in this transient period.

The tundra landscape is characterized by a mosaic of wet, dry and mesic areas, which forms in response to variations in topography, soils and hydrology. A basic catena on slopes, which could be affected by slope shape and by local geomorphology such as landforms of patterned ground (e.g. Klimowicz and Uziak, 1996; Bockheim et al., 2000, Macdonald et al., 1999) may give insight into the influence of different factors (e.g. topography, vegetation) on soil processes. In Alpine areas of Northern Italy and Switzerland, the relationship between climate and pedological processes is often non-linear and is thought to be overshadowed by the pronounced podzolisation effect below and near the timberline (1400–1900 m asl) (Mirabella and Sartori, 1998; Mirabella et al., 2002). Element leaching and weathering rates are greatest in the subalpine forest range, while lower weathering rates have been measured at both very high and low altitudes (Egli et al., 2004). Temperature and organic matter quality, connected to the type of vegetation cover, have been postulated as the primary drivers of nitrogen flux in tundra soils.

Burns and Tonkin (1982) proposed a Synthetic Alpine Slope model in which are described different topographic elements that explain relative winter snow cover, vegetation and hence soil distribution. Pedogenesis along environmental gradients has been described for several sites near timberline (e.g. Stanton et al., 1994). Patterned ground could affect the catena, but it is also correlated with slope position and shape (e.g. Butler and Malanson, 1999).

In this work, we aimed to evaluate pedogenetic processes and plant distribution along an elevation gradient in the alpine tundra in the western Italian Alps. Additionally, we measured nitrogen pools, including biomass N, in soils as a proxy for biological activity. We then used these data to (1) identify main pedogenetic processes along the gradient and main environmental parameters controlling pedogenesis, (2) test the ability of soils to conserve nutrients at various pedogenetic degrees, and (3) evaluate how the plant community respond to the various nutrient status in tundra soils.

MATERIAL AND METHODS

The study area (LTER site "High elevation areas in the Northwestern Alps") for this investigation is located in North West Italy, close to the Monte Rosa

Massif (4634 m asl) (Figure 1). Meteorological data are continuously recorded from 2005 by an automatic weather station of the Italian Army (Comando Truppe Alpine-Servizio Meteomont), located at an elevation of 2901 m asl. The area is characterized by a precipitation maximum during autumn, with an average cumulative annual snowfall equal to 900 cm. The mean annual air temperature is equal to -4.1 °C. The bedrock is constituted mainly by micaschists, with some inclusions of ophiolites and calcicschists. Soil profiles were dug and described at 5 sites along an elevation gradient from 2525 m asl to 2840 m asl, in the upper part of a glacial valley(Figure 2).

Soil material was collected from excavated profile pits (late summer 2008). Due to the stony character of the soils, around 4 to 5 kg of soil material was collected per soil horizon. In order to yield reasonable results, large soil sampling volumes are needed for soils in alpine areas (Hitz et al., 2002; Egli et al., 2003). Soil bulk density (fine earth + stony material) (BD) was determined by a specific soil core sampler. Taking advantage of the profile pits, undisturbed soil samples were taken down to the C horizon. Genetic horizons were identified and physical properties described following standard methodology (Soil Survey Division Staff, 1993).

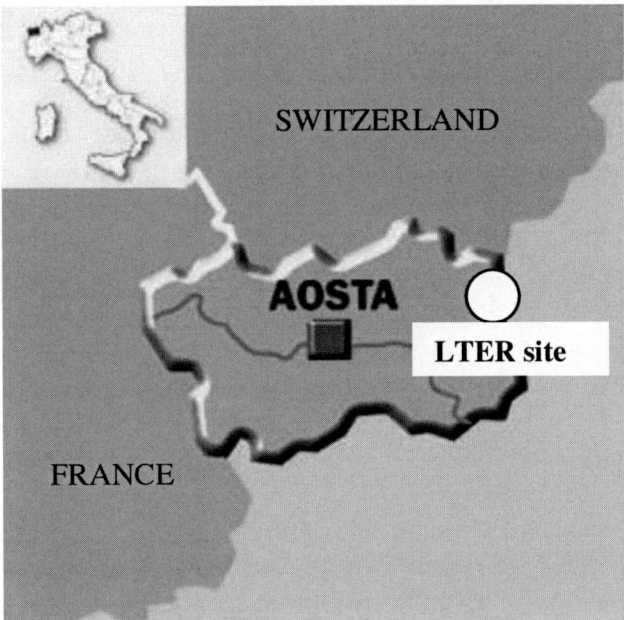

Figure 1. Location of the study site.

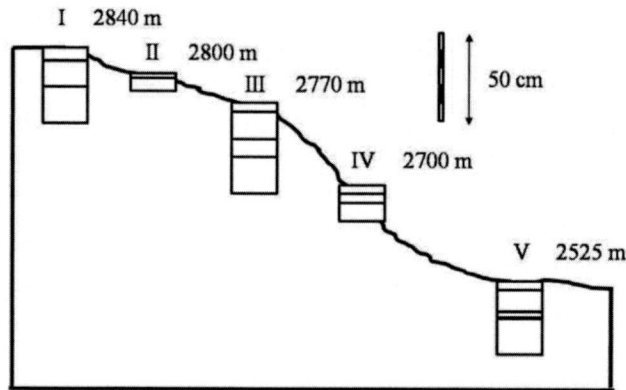

Figure 2. Location of the 5 sampling sites along the elevation gradient. Elevation, horizonation and horizon depths are shown (note that the distance between sites and the elevation gradient are not in scale).

In the laboratory, samples were dried and passed through a 2-mm sieve. Soil chemistry (pH, electrical conductivity (EC), cation exchange capacity (CEC), exchangeable cations (Ca exc, Mg exc, K exc) and base saturation (BS)) and particle size distribution (Clay, Fine silt (F. silt), Coarse silt (C. silt), Fine sand (F. sand), Coarse sand (C. sand), and Skeleton (Skel)) were determined following standard methods (SISS, 1998, 2000). Total C (TOC) and N (TN) contents of the soil were measured with a C/H/N analyser (Elementar Vario EL).

C pools were calculated in the upper mineral horizons (A) and in the lower mineral horizons (C, AC, BC), using measured bulk densities per unit area basis. N pools were calculated in the topsoil (0-10 cm depth). Additionally, fresh topsoil (0-10 cm depth) samples were taken in triplicate for each site in October 2008. These soil samples were extracted with 0.5M K_2SO_4 (1: 5 weight: volume) within 24 hours of returning from the field. Before extraction, each sample was divided in two subsamples, one was extracted and one fumigated for 18 hours with chloroform and subsequently extracted as explained before. On soil extracts ammonium-N (N-NH_4^+), nitrate-N (N-NO_3^-), dissolved organic nitrogen (DON) and microbial N (N_{micr}) were determined following previously published protocols (Crooke and Simpson, 1971; Brookes et al., 1985).

Data on the vegetation structure were collected close to each soil profile, covering a surface of 16 m^2; each sampling site has been further divided into 4 sub-areas of 4 m^2. The abundances of species were recorded as cover percentages. Nomenclature followed Tutin et al. (1993) for the vascular plants, Cortini Pedrotti (2001-2006) for mosses and **Schumacker and Vaňa (2000) for liverworts and hornworts**. A percentile tabulation of the plants community was calculated by the geographycal distribution and the life form classes according to Raunkiaer's classification. Ecological values (Landolt, 1977) were calculated as means weighted by frequencies.

Statistical Analysis

Vegetation data for all species were transformed as ln (x+1) to stabilise variance. The environmental variables were also transformed to adapt them for statistical treatment. To evaluate the population structure and to express sample relationships in ecological space we performed Non Metric multidimensional Scaling (NMDS) (Kruskal, 1964; Mather, 1976). NMDS was chosen to order plots in species space and to graphically underline the interrelationships between communities and their relationship to environmental variables. The method was ideal in our context because of its capacity to find the lowest stress between the ranked distances in the original multidimensional space and the ranked distances in the reduced ordination space by an iterative search; in this way NMDS offered the best graphic representation of the data set. Sørensen distances were used. By means of a joint plot, we superimposed the separate environmental variables on the ordination resulting from NMDS (Peterson and McCune, 2001).

All statistical and multivariate analyses were performed using SPSS for Windows version 12.0, and PC-ORD (McCune and Mefford, 1999).

RESULTS

Soil Characteristics

Soils were classified according to the WRB (IUSS Working Group WRB, 2006) as reported in Table 1. Soil depth in the five study areas ranged from 10 to more than 50 cm. Soil profiles exhibited a fairly consistent horizonation. All of the profiles had a black to dark brown A horizon, in some cases underlying thin Oi/Oe horizons. Structure was usually weakly developed fine granular, except at lowest elevation, where it was moderately developed subangolar blocky. The soil matrix revealed a prevalence of the 2.5Y 4/3 and 2.5Y 5/3 colours (Table 1). Texture was generally coarse, ranging from sandy to loamy sand, with a clay content lower than 30 gkg^{-1} (Table 2), with most values around 15 gkg^{-1}. The percentage of skeleton ranged between 3.0 and 66.3 %, with lower values in the topsoil only in sites III and V. The bulk density ranged between 826 and 1590 kgm^{-3}, with values higher at sites I and II. The EC ranged between 19.7 and 165.5 µScm^{-1}, with higher values in the topsoil at all sites. The soil pH ranged between 4.4 and 6.1 at all sites except for C horizon at site I (7.2). The pH was lower in the upper horizons at all sites except site IV, where it was quite homogeneous along the profile. The organic carbon concentration ranged between 4.0 and 75.0 gkg^{-1}, with higher values in the topsoil in all sites (Table 3). The same trend was found also for the total nitrogen content (TN). The C pools in the upper mineral horizons (A) ranged between 0.9 kgCm^{-2} in site II and 9.0 kgCm^{-2} in site III (Table 4). In the deeper horizons the C pools ranged between 0.3 and 4.5 kgCm^{-2}.

At all sites Ca was the dominant exchangeable base cation, with higher values at site IV. The base saturation (BS) at this site ranged between 38 and 53%, while at all the other sites BS was considerably lower, with maximum values of 19%.

TN pools in the topsoil ranged 0.15-0.34 kgm^{-2}, with highest values in site III. Inorganic nitrogen (IN) pool was less than 1% of the TN pool, with ammonium (N-NH$_4^+$) as the most important IN form. Dissolved organic nitrogen (DON) accounted for another 1% of the total N, and N associated with microbial biomass (N micr) was about 2% of TN (Table 5). We defined labile N as the sum of total dissolved N and the N micr. This pool accounted for slightly more than 3% on TN. Therefore, most (97%) of the TN is constituted by recalcitrant, non-labile organic nitrogen.

The C/N ratio ranged between 10 and 21, with highest values in site III. The pH was significantly correlated with the organic carbon (TOC) (r=-0.529, p<0.05). The C/N ratio was significantly correlated with the slope angle (r=-0.576, p<0.05) but not with elevation (Table 6).

The cation exchange capacity ranged between 4.78 and 22.89 meq100g^{-1}, with values higher in the topsoil only in sites IV and V. The cation exchange capacity was significantly correlated with the carbon content (TOC) (r=0.858, p<0.01) (Table 6).

In the topsoil N-NH$_4^+$ concentration ranged between 3.1 and 13.2 mgNkg^{-1}, with significantly lower values at sites I and IV than at other sites (Figure 3). The nitrate (N-NO$_3^-$) concentration showed the same pattern, with values approximately one order of magnitude lower than ammonium concentrations in all sites. Ammonium and nitrate concentration were significantly correlated with the elevation (r=-0.516, p<0.05; r=-0.691, p<0.01) and the soil humidity (r=0.737; r=0.804, p<0.01) (Table 7).

Table 1. Some characteristics of the five investigated sites.

Soil	Elevation m asl	Slope °	Horizon	Depth cm	Skeleton %	BD kgm^{-3}	Colour (air dried)
I Leptic Regosol Dystric	2840	0	A	0-7	16.3	1590	2.5Y 4/2
			C1	7-20	27.0	1564	2.5Y 5/3
			C2	20-40	16.0	1436	2.5Y 5/3
II Lithic Leptosol Dystric	2800	0	A	0-3	30.4	1167	10YR 3/2
			C	3-10	14.6	1167	2.5Y 4/3
III Haplic Umbrisol	2770	6	A1	0-5	11.0	1023	10YR 3/2
			A2	5-20/25	16.3	891	2.5Y 5/3
			A3	20/25-30	18.9	940	2.5Y 4/4
			AC	30-50+	30.6	874	2.5Y 4/3
IV Haplic Leptosol	2703	37	A1	0-5	9.9	826	2.5Y 4/3
			A2	5-10	9.7	826	2.5Y 4/3
			AC	10-20	3.0	1040	2.5Y 5/4
V Umbrisol Brunic Skeletic	2525	22	A1	0-5	50.5	1026	2.5Y 4/3
			A2	5-20	54.2	1026	2.5Y 5/3
			Bw	20-25	66.3	1030	2.5Y 5/4
			BC	25-40	53.5	1030	2.5Y 6/3

Table 2. Some physical and chemical properties of the investigated soils.

Soil	Horizon	clay	fine silt	coarse silt	fine sand	coarse sand	EC	pH
		gkg^{-1}					□ Scm^{-1}	
I	A	10.4	43.0	91.5	332.5	522.5	45.5	4.6
	C1	11.9	69.0	103.5	319.5	496.0	13.2	5.4
	C2	12.4	65.8	102.0	325.7	494.0	19.7	7.2
II	A	14.1	57.5	158.4	311.5	458.5	146.3	4.4
	C	17.7	139.0	172.0	328.8	342.5	18.2	4.7
III	A1	14.2	39.0	77.8	219.0	650.0	99.7	4.5
	A2	24.5	98.0	119.7	339.3	418.5	51.5	4.4
	A3	16.7	72.5	71.5	288.8	550.5	50.1	4.6
	AC	16.2	63.8	107.5	258.5	554.0	21.4	5.0
IV	A1	11.8	88.9	125.5	297.2	476.5	165.5	5.6
	A2	16.3	94.8	121.9	266.5	500.5	124.6	5.4

	AC	13.2	115.7	193.7	369.3	308.0	60.5	5.5
V	A1	15.6	64.4	80.4	177.0	662.5	160.4	5.0
	A2	13.4	63.5	77.9	272.6	572.5	66.8	5.2
	Bw	19.0	103.4	102.4	259.8	515.5	33.7	5.3
	BC	28.0	111.5	90.0	177.1	593.5	36.7	6.1

The microbial N concentration was extremely variable, with values significantly lower at site I (Figure 4). The microbial N concentration was significantly correlated with the elevation (r=-0.735, p<0.01) and the water content (r=0.858, p<0.01) (Table 7). The DON concentration was lower than N associated with microbial biomass at all sites, with significantly lower values in site I (Figure 4).

Table 3. Some physical and chemical properties of the investigated soils.

Soil	Horizon	TOC	TN	C/N	CEC	Ca exc	Mg exc	K exc	BS
		gkg^{-1}			meq100g^{-1}				%
I	A	18.4	1.4	13	7.84	0.43	0.10	0.08	7.8
	C1	6.5	0.5	14	4.78	0.08	0.01	0.01	2.1
	C2	14.9	0.8	18	8.82	0.11	0.01	0.01	1.5
II	A	37.9	3.0	13	13.12	0.70	0.13	0.15	7.5
	C	26.2	1.8	14	14.32	0.17	0.02	0.02	1.5
III	A1	75.0	5.1	15	22.00	0.94	0.24	0.27	6.6
	A2	53.1	2.6	20	22.89	0.26	0.04	0.02	1.4
	A3	39.8	1.9	21	22.22	0.17	0.02	0.01	0.9
	AC	35.0	1.9	18	21.60	0.22	0.02	0.01	1.1
IV	A1	41.5	3.4	12	18.10	8.16	1.23	0.22	53.1
	A2	40.1	3.5	12	17.11	6.78	1.08	0.19	47.0
	AC	20.9	2.0	10	13.75	4.51	0.69	0.09	38.4
V	A1	66.1	4.7	14	20.83	3.59	2.88	0.30	32.5
	A2	21.4	1.8	12	11.77	1.10	1.08	0.05	19.0
	Bw	12.3	1.0	12	12.31	0.89	0.80	0.02	13.8
	BC	4.0	0.4	10	5.09	0.46	0.45	0.01	17.9

Table 4. Estimates of soil C pools per unit land area at the different sites.

Horizon	A		Bw		C, AC, BC		
Sites	depth cm	Tot pool kgCm^{-2}	depth cm	Tot pool kgCm^{-2}	depth cm	Tot pool kgCm^{-2}	C in C-horizon % of Tot pool
I	7	1.7			30	4.5	72
II	3	0.9			7	1.8	66
III	20	9.0			20	4.2	32
IV	10	3.0			10	2.1	41
V	17	3.6	8	0.3	15	0.3	6.7

Table 5. TN stocks and N forms stocks (kg ha^{-1}) in topsoils (0-10 cm) at the five sites. N forms are also reported as % on TN in parentheses.

Site	TN	N-NH$_4^+$	N-NO$_3^-$	DON	TDN	N micr	Labile N[*]
	kg ha^{-1}			kg ha^{-1} (% on TN)			
I	1500	4.6 (0.3)	0.3 (0.0)	15.1 (1.0)	19.9 (1.3)	22.1 (1.4)	42.0 (2.8)
II	1800	8.8 (0.5)	1.1 (0.1)	14.0 (0.8)	23.9 (1.3)	40.1 (2.2)	64.0 (3.5)
III	3400	11.4 (0.3)	3.0 (0.1)	21.3 (0.6)	35.9 (1.1)	73.8 (2.2)	109.7 (3.2)
IV	2500	4.1 (0.2)	0.5 (0.0)	11.1 (0.4)	15.6 (0.6)	45.0 (1.8)	60.6 (2.4)
V	1700	7.1 (0.4)	2.1 (0.1)	10.4 (0.6)	19.6 (1.2)	52.4 (3.1)	72.0 (4.2)

(*) Labile N pool was calculated by summing TDN and microbial N pool

Table 6. Correlation matrix between soil properties (*p<0.05; **p<0.01).

Profiles	Slope	Elev.	Skel.	pH	EC	C	N	C/N	CEC	Ca exc	Mg exc	K exc	BS	Clay	F. silt	C. silt	F. sand
Slope																	
Elev.	-0.629**																
Skel.	0.065	0.765**															
pH	0.254	-0.147	0.087														
EC	0.471	-0.237	-0.079	-0.223													
C	0.032	0.044	-0.268	-0.529*	0.630**												
N	0.244	-0.067	-0.253	-0.453	0.803**	0.939**											
C/N	-0.576*	0.486	-0.250	-0.160	-0.299	0.330	0.013										
CEC	0.111	0.024	-0.261	-0.513*	0.399	0.858**	0.726**	0.521*									
Ca exc	0.841**	-0.234	-0.298	0.155	0.703**	0.257	0.479	-0.436	0.235								
Mg exc	0.641**	-0.698**	0.366	0.071	0.659**	0.345	0.510*	-0.385	0.211	0.601*							
K exc	0.388	-0.183	-0.148	-0.219	0.890**	0.738**	0.906**	-0.296	0.417	0.625**	0.669**						
BS	0.927**	-0.449	-0.098	0.216	0.673**	0.138	0.393	-0.606*	0.094	0.960**	0.701**	0.593*					
Clay	0.070	-0.409	0.366	-0.040	-0.220	-0.041	-0.177	0.084	0.087	-0.242	-0.067	-0.303	-0.147				
F. silt	0.381	-0.244	-0.014	0.161	-0.230	-0.331	-0.295	-0.229	-0.092	0.200	0.004	-0.327	0.242	0.526*			
C. silt	0.166	0.272	-0.436	-0.049	0.042	-0.152	-0.033	-0.268	-0.061	0.266	-0.164	-0.062	0.219	-0.072	0.588*		
F. sand	-0.220	0.655**	-0.598*	-0.045	-0.293	-0.293	-0.319	0.191	-0.148	-0.040	-0.519*	-0.404	-0.159	-0.391	0.175	0.606*	
C. sand	-0.046	-0.393	0.497	0.000	0.233	0.325	0.293	0.050	0.131	-0.121	0.366	0.368	-0.051	0.055	-0.634**	-0.890**	-0.840**

Table 7. Correlation matrix between selected chemical properties of fresh topsoil samples collected at the five sites ($*p<0.05$; $**p<0.01$).

Topsoils	Elevation	Slope	WC	pH	EC	$N-NH_4^+$	$N-NO_3^-$	TDN	DON
Elevation									
Slope	-0.639*								
WC	-0.905**	0.575*							
pH	-0.686**	0.702**	0.782**						
EC	0.534*	-0.494	-0.509	-0.774**					
$N-NH_4^+$	-0.516*	-0.047	0.737**	0.487	-0.257				
$N-NO_3^-$	-0.691**	0.184	0.804**	0.636*	-0.500	0.877**			
TDN	-0.473	0.028	0.700**	0.554*	-0.395	0.887**	0.816**		
DON	-0.289	0.067	0.501	0.473	-0.386	0.630*	0.555*	0.911**	
N_{micr}	-0.735**	0.400	0.858**	0.724**	-0.487	0.777**	0.803**	0.621*	0.330

Abbreviations: WC water content

Table 8. Vascular species and their abundance (%). The value of 0.1 means that only a single individual was present. Plots are in roman numbers, subplots are in italics.

	Ia	Ib	Ic	Id	IIa	IIb	IIc	IId	IIIa	IIIb	IIIc	IIId	IVa	IVb	IVc	IVd	Va	Vb	Vc	Vd
Alchemilla vulgaris group	-	-	-	-	-	-	-	-	-	-	-	-	-	-	0.5	-	-	-	-	-
Agrostis agrostiflora (G. Beck) Rauschert	-	-	-	-	-	-	-	-	-	-	-	-	6	4	2	2	-	-	-	-
Agrostis alpina Scop.	-	-	-	-	-	-	-	-	-	-	-	-	-	-	-	-	-	0.1	-	-
Agrostis rupestris All.	-	-	-	-	-	-	-	-	0.5	-	0.5	-	0.5	0.5	1	0.5	-	-	-	-
Alchemilla fissa Günther et Schummel	-	-	-	-	-	-	-	-	-	-	-	-	2	0.5	1	0.5	-	-	-	-
Alchemilla pentaphyllea L.	-	-	-	-	-	-	-	-	-	-	-	-	0.5	0.5	0.5	-	-	-	-	-
Antennaria carpatica (Wahlenb.) Bluff et Fingerh.	-	-	-	-	-	-	-	-	-	0.5	-	-	-	-	-	-	-	-	-	-
Anthoxanthum odoratum L.	-	-	-	-	-	-	-	-	-	-	-	-	-	-	0.5	0.5	0.5	0.1	0.5	0.5
Armeria maritima (Miller) Willd.	-	-	-	-	-	-	-	-	-	-	-	-	0.5	-	0.5	0.5	0.1	0.1	0.1	0.1
Bartsia alpina L.	-	-	-	-	-	-	-	-	-	-	-	-	0.5	0.5	0.5	0.5	0.5	0.1	0.5	0.5
Campanula scheuchzeri Vill.	-	0.5	-	-	-	-	-	-	0.5	-	-	-	-	-	-	-	-	-	-	-
Cardamine resedifolia L.	-	-	-	0.5	-	-	-	-	-	-	-	-	0.5	-	0.5	0.5	-	-	-	-
Carex atrata L.	-	-	-	-	-	-	-	-	-	-	-	-	-	-	0.5	-	-	-	-	-
Carex curvula All.	-	-	-	-	-	1	-	-	15	3	40	-	-	2	0.5	1	4	3	4	4
Carex sempervirens Vill.	-	-	-	-	-	-	-	-	-	-	8	-	3	1	2	1	-	-	-	-
Cerastium arvense L.	-	-	-	-	-	-	-	-	-	-	-	0.5	0.5	-	0.5	0.5	-	-	-	-
Cerastium cerastoides (L.) Britton	0.1	-	-	-	-	-	-	-	-	-	-	-	-	-	-	-	-	-	-	-
Cerastium uniflorum Clairv.	-	0.5	-	0.5	-	-	-	-	-	-	-	-	-	-	-	-	-	-	-	-
Cirsium spinosissimum (L.) Scop.	-	-	-	-	-	-	-	-	-	-	-	-	5	-	5	5	-	-	-	-
Euphrasia minima Jacq. ex DC. in Lam. Et DC.	-	0.5	-	-	0.5	-	-	-	0.5	0.5	0.5	0.5	0.5	1	0.5	1	0.5	1	0.5	0.1
Festuca halleri All.	-	0.5	0.5	-	-	-	-	-	1	0.5	-	0.5	-	-	0.5	0.5	0.1	0.1	0.1	0.1
Festuca violacea Schleicher ex Gaudin	-	-	-	-	-	-	-	-	3	1	-	-	0.5	0.5	1	0.5	0.5	0.5	1	0.5
Galium megalospermum All.	-	-	-	-	-	-	-	-	-	-	-	-	-	-	1	-	-	-	-	-
Gentiana acaulis L.	-	-	-	-	-	-	-	-	-	-	-	-	-	-	-	0.5	-	-	-	-
Gentiana bavarica L.	0.5	0.5	-	0.5	-	-	-	-	-	-	-	-	-	-	-	-	-	-	-	-
Gentiana nivalis L.	-	-	-	-	-	-	-	-	-	-	-	-	-	-	-	-	0.1	0.1	0.1	0.1
Geum montanum L.	-	-	-	-	-	-	-	-	-	-	-	-	3	1	2	1	-	0.1	0.1	-
Gnaphalium supinum L.	3	3	2	5	0.1	-	0.1	-	-	-	-	-	-	-	-	-	-	-	-	-
Hieracium glanduliferum Hoppe in Sturm	-	-	-	-	-	-	-	-	1	3	-	0.5	-	-	-	-	-	0.1	-	-
Hieracium villosum Jacq.	-	-	-	-	-	-	-	-	-	-	-	-	-	-	-	-	0.5	1	1	2
Homogyne alpina (L.) Cass.	-	-	-	-	-	-	-	-	1	-	-	-	-	-	-	0.5	0.5	0.5	0.1	0.1
Juncus trifidus L.	-	-	-	-	-	-	-	-	-	1	-	-	-	-	-	-	-	-	-	-
Kobresia myosuroides (Vill.) Fiori et Paol.	-	-	-	-	-	-	-	-	0.5	-	-	-	-	-	-	-	-	-	-	-
Leontodon pyrenaicus Gouan	-	-	-	0.5	0.5	-	-	-	0.5	-	0.5	-	-	-	0.5	-	-	-	-	-
Leucanthemopsis alpina (L.) Heywood	1	1	-	2	0.5	0.5	-	0.5	0.5	-	-	-	0.5	1	1	1	0.5	0.5	0.5	0.5
Ligusticum mutellina (L.) Crantz	-	-	-	-	-	-	-	-	-	-	-	-	-	0.5	-	-	-	0.1	0.1	0.1
Linum catharticum L.	-	-	-	-	-	-	-	-	-	-	-	-	0.5	-	1	1	-	-	0.1	-

Table 8. (Continued)

Species	1	2	3	4	5	6	7	8	9	10	11	12
Lotus alpinus (DC.) Schleicher ex Ramond	-	-	-	-	-	-	-	-	-	-	0.5	0.1
Luzula alpinopilosa (Chaix) Breistr.	2	-	-	-	-	-	-	-	-	1	-	-
Minuartia sedoides (L.) Hiern	3	-	0.5	-	0.5	-	-	-	-	-	-	-
Myosotis alpestris F. W. Schmidt	-	-	-	-	-	-	0.5	-	-	-	0.1	0.1
Pedicularis kerneri Dalla Torre	-	-	-	-	0.5	-	-	-	-	-	0.1	-
Phyteuma globularifolium Stemb. et Hoppe	-	-	-	-	0.5	0.1	1	0.5	0.5	-	0.1	-
Phyteuma hemisphaericum L.	-	0.5	-	0.5	0.5	-	-	0.5	0.5	-	0.1	-
Poa alpina L.	5	1	3	0.5	-	-	0.5	0.5	0.5	-	-	-
Poa laxa Haenke in J. Jirasek	-	-	-	-	10	3	10	1	5	2	2	0.5	0.5
Polygonum viviparum L.	-	-	-	-	-	-	1	5	2	2	0.1	0.1
Potentilla aurea L.	-	-	-	-	0.5	-	-	-	-	-	-	-
Primula hirsuta All.	-	-	0.5	-	-	-	0.5	-	-	-	-	-
Ranunculus glacialis L.	0.5	-	-	-	1	-	0.5	0.5	1	0.5	1	0.5
Ranunculus grenierianus Jordan in F. W. Schultz	-	-	-	-	0.5	1	1	0.5	0.5	-	0.1	0.1
Salix herbacea L.	-	15	3	25	5	10	-	0.5	-	0.5	-	-
Saxifraga bryoides L.	-	-	0.5	-	-	1	-	-	-	-	-	-
Saxifraga retusa Gouan	-	-	-	-	-	0.5	-	-	-	-	-	-
Sedum alpestre Vill.	0.5	0.5	-	-	-	-	0.5	-	-	-	-	-
Senecio halleri Dandy	-	-	0.1	-	0.1	0.5	-	0.5	-	-	-	-
Silene acaulis (L.) Jacq.	-	-	5	-	6	1	3	2	-	-	-	-
Soldanella alpina L.	-	-	-	0.5	1	10	4	-	0.5	-	0.5	-
Taraxacum alpinum group	-	-	0.5	-	-	-	1	3	-	0.5	1	0.5
Vaccinium myrtillus L.	-	-	-	-	-	-	-	-	-	1	1	2
Vaccinium uliginosum L.	-	-	-	1	5	3	-	-	-	5	5	4
Valeriana celtica L.	-	-	-	0.5	-	-	-	-	-	-	2	-
Veronica alpina L.	0.5	0.5	0.1	-	-	-	0.5	0.5	0.5	-	-	0.1
Viola biflora L.	-	-	-	-	-	-	-	-	-	-	-	-

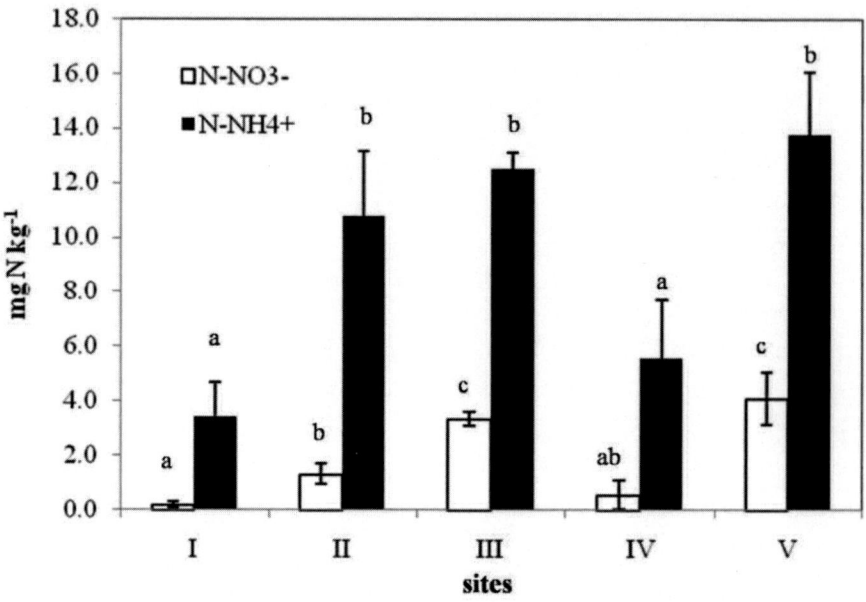

Figure 3. Inorganic nitrogen concentration in topsoil (0-10 cm, n=3, error bars are standard deviations) at the five sites. Different letters indicate significant differences between sites ($p<0.05$).

Plant Characteristics

A total of 64 vascular species were found at the sampling sites (Table 8). The life form spectrum (Figure 5a) showed that Hemicryptophytes, with buds at or near the soil surface were dominant (91.07%), followed by chamaephytes (6.82%), geophytes (1.21%) and therophytes (0.91%). From the phytogeographical point of view, as shown in the chorological spectrum (Figure 5b), the species were mainly south european mountain (49.44%), artic alpine (29.74%), west alpine (7.81%) and alpine (6.82%) while the ones with wider range, as european or eurosibiric, were less frequent (6.19%).

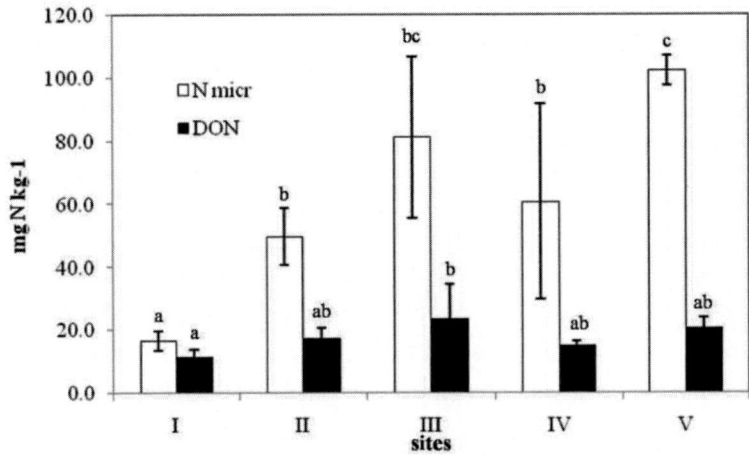

Figure 4. Dissolved organic nitrogen and microbial N concentrations in topsoil (0-10cm, n=3, error bars are standard deviations) at the five sites. Different letters indicate significant differences between sites (p<0.05).

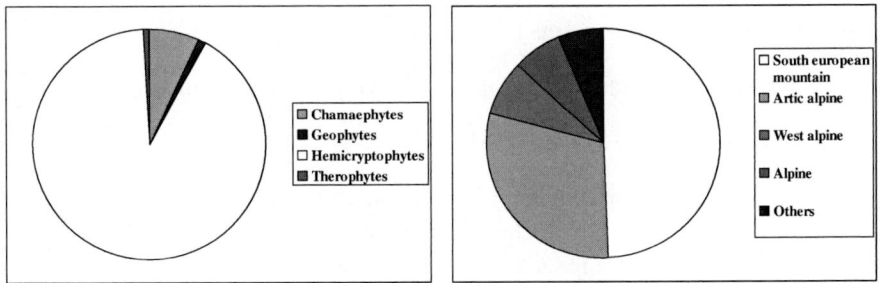

Figure 5. Biological (a) and chorological (b) spectrum of the flora

Landolt ecological spectrum values (Figure 6) suggest an heterogeneous environment where parameters such as light (L) and temperature (T) display sharp changes along the elevation gradient and/or as function of morphological features at single sites. Consequently, the vegetation physiognomy changes as a function of altitude and morphology. Sites I and II, characterized by relatively low slope angles, are covered by a nival vegetation, with typical species such as *Poa laxa, Minuartia sedoides, Leucanthemopsis alpina,* together with species that can tolerate long-lasting snow cover, such as *Salix herbacea* and *Gnaphalium supinum* and typical bryophytes of the snowbed areas.

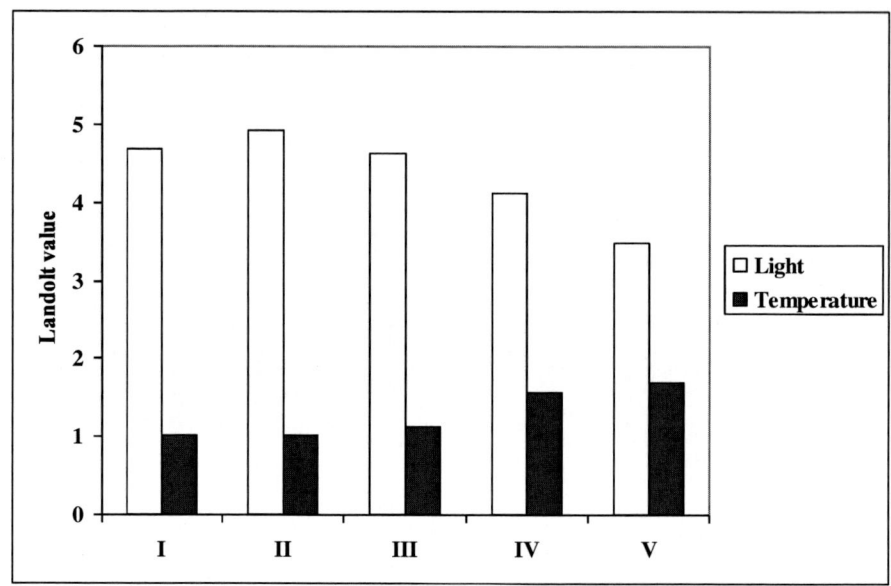

Figure 6. Ecological spectrum of the flora for light and temperature (Landolt values).

Site III has a considerably higher vegetation cover, compared to previous sites, with most of species being characteristic of *Carex curvula-* dominated alpine grasslands.

At site IV, characterized by a steep slope, there is evidence of snowbed species, similarly as at higher altitude, and grassland species such as *Agrostis agrostiflora* and *Carex sempervirens*, together with several species found at all other sites. Despite the steep slope, there is still evidence of species associated with long-lasting snow cover, such as *Alchemilla pentaphyllea*.

At site V, despite the lower steepness, the plant community is quite comparable with the one found at site IV. For the first time along the elevation gradient we detected *Vaccinium myrtillus*, together with *V. uliginosum*. Even so, the woody species are not predominant in the plant community. The ordination analysis clearly indicates the distinct nature of the alpine tundra subplots, which form well separate group corresponding to the five plots.

The NMDS autopilot mode selected 2 axes as the best representation in the space of ordination. Axes explained 80.4% of data variance. After rotation, the first axis accounted for 54.5% of the variation in community. Figure 7 showed in NMDS space the position of all plots that were enclosed by lines to identify the 5 groups of subplots. These five groups were also well separated, thus confirming the results of the former evaluations. The position of both subplots and sites was explained by slope and nine soil variables. The horizontal axis, accounting for a larger variance, openly described the variability of sites and/or subplots in terms of a single and dominant gradient defined by nitrogen forms, humidity and organic matter content. Both subplots and sites were also separated along the vertical axis that explained 25.9% of the total species variability. In this case, from bottom to top, pH values and exchangeable Mg^{2+} content were related with the intensification of the slope.

DISCUSSION

Pedogenetic Processes

The major pedogenetic processes responsible for soil formation in this environment are: a) organic matter accumulation and melanization; b) cryoturbation; c) erosion and deposition of material. Melanization appeared as a sharp contrast in darkness between the surface horizons and deeper parts of the solum. The accumulation of organic matter is encouraged by low temperatures that reduce rates of decomposition. The presence of irregular horizon boundaries within the profiles are evidence of cryoturbation, as these soils are frost affected, mainly in early winter and late spring, when the reduced snow cover may not be sufficient to insulate the soil from the air temperature. During winter, the thick snow cover that generally characterize these sites insulate the soil, which maintain a temperature close to 0°C, independently by air temperature (Freppaz et al., 2008a). Air temperatures are below 0°C for more than half the year at these elevations. The insulation of the snowpack against these temperatures was generally remarkable: the lowest soil temperature recorded over 5 years was -7°C (27th January 2006, snow depth 70 cm, air temperature -22°C). These data are higher than those reported by Ley et al. (2004), in a talus soil at 3750 m in the Colorado Rocky Mountains (-2.9 °C).

The investigated soils usually have a high skeleton proportion (material >2 mm diameter; Table 1); this is typical for soils on alpine moraines (Egli et al., 2003). All soils of the investigation areas are, furthermore, characterised by a sandy to sandy–silty texture.

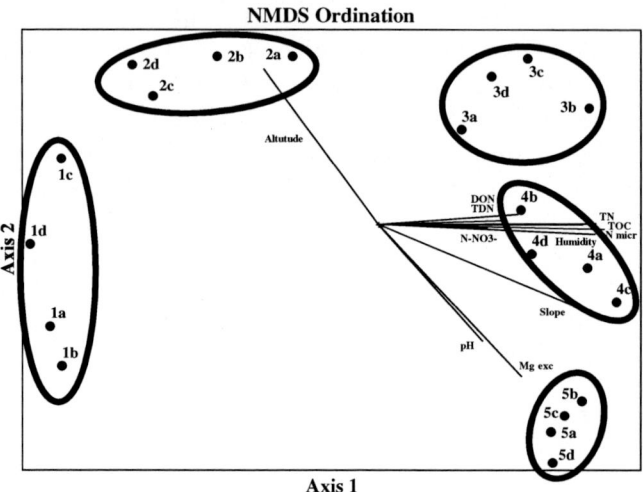

Figure 7. Results of NMDS for all plots. Only variables at a cut-off Kendall τ >0.40 for joint plots are shown.

C stocks in the upper horizons don't increase with decreasing altitude (maximum in site III), thus indicating that accumulation of organic matter is not only controlled by the soil thermal regime. At higher elevation the organic matter in A horizon per unit land area reaches the minimum in site II, while the C concentration per unit topsoil dry matter reaches the minimum at site I. This is in contrast with what reported by Korner (2003), who, disregarding special situations, reports that the C concentration remains high at higher elevations. The reduction of SOM pools in the upper alpine zone is a consequence of reduced plant cover, rooted soil depth and reduced plant productivity. The soil carbon pools found in A horizons, taking into account the horizon depths, in this study are comprised between 0.9 e 9.0 kgm^{-2}, values slightly lower than those reported (4-22 kgm^{-2}) for closed alpine vegetation at medium alpine altitudes, which is also the range found in crop fields, pastures and forests at low elevations (Körner, 2003). Körner et al. (1996) in a closed turf of Carex curvula in the Swiss Alps at an elevation of 2470 m asl reported a value of 9 $kgCm^{-2}$. Since annual net primary production is 5 to 10 times lower in alpine compared with lowland ecosystems, the similar range of soil C pools found under closed alpine vegetation is remarkable. In alpine ecosystems most of the C pool is contained in the soil organic matter, characterized by a high residence time compared with life biomass.

As expected, C concentration usually decreases with increasing depth in the soil profile; however the contrast between C in the surface horizons and in the deeper horizons becomes sharper with decreasing altitude, suggesting that organic matter accumulation is higher at lower sites.

At sites III, IV and V the C stocks in C, AC and BC horizons is 2-10 times lower than in A horizons. At higher elevation, we found an opposite trend, with higher carbon accumulation in deeper horizons, with values at sites I and II up to 3 times higher than A horizons. The % contribution of C in deeper horizons range between 6.7 and 72%, values in accordance with Körner (2003), who, in mature alpine soils found that from 6 to 32% of the total C is accreted in the C horizon at depths of between 0.3 and 1 m.

CEC and Base saturation both increase toward the decreasing elevation gradient. This pattern may be connected with the litology in the lower sites, characterized by calcschists (IV) and ophiolites (V). Additionally, a higher base saturation in the lower profiles indicate a better capacity to retain nutrients, as confirmed by the N patterns, that will be discussed later.

The pedogenesis seems enhanced at lower elevation, where B horizons were found. A strong role in controlling pedogenetic processes is played by soil erosion. Soil erosion and deposition may occur under steep slopes, determining a loss and a subsequent redeposition of soil material, and the magnitude and the effects of such process are evident at site IV: the depth of the solum is considerably reduced compared to site III and V, a B_w horizon was not detected and soil structure was weakly developed compared to both III and V. The strong implication on the availability and retention capacity of nutrients will be discussed later.

Soil Nutrient Pools

The TOC concentration in the upper profiles of the investigated sites was slightly lower than that reported by Makarov et al. (2003) for alpine ecosystems in northern Caucasus, at comparable elevation. Surprisingly, TOC concentrations at our sites were comparable with those reported for organic layers in forested ecosystems of the Eastern Alps (Egli et al., 2009). The total nitrogen in these soils is correlated with soil organic matter concentrations (Körner, 2003). TN content was in the range reported by Makarov et al. (2003) for Alpine Caucasian Ecosystems and Brooks et al. (1996), and Fisk and Schmidt (1995) for alpine tundra at Niwot Ridge. Total N pools in the topsoil of these sites ranged ranged between 0.15 and 0.34 $kgNm^{-2}$, values slightly lower in comparison to what found for low altitude ecosystems (a range of 0.75 to 1 kgNm^{-2}, with an extreme of 2 for coniferous forests, compiled by Killham (1994).

N labile compounds (i.e. ammonium, nitrate, DON, microbial N) are known to display a strong seasonality in both alpine (Brooks and Williams, 1999) and subalpine (Filippa et al., 2009; Neff et al., 1994) ecosystems. In this study, soil samples for labile N forms were collected in autumn; therefore we compared our data with available data from the end of the growing season. Ammonium and nitrate concentrations at our sites were comparable with those reported by Fisk and Schmidt (1995) e Makarov et al. (2003) for other alpine ecosystems. Somewhat more surprisingly, our data are comparable to those reported for subalpine meadow ecosystems (Filippa et al., 2009), but slightly lower than those reported in the same area in subalpine forest and meadow ecosystems (Freppaz et al., 2008b). In a similar fashion, comparisons for DON and microbial N concentrations are consistent with those of ammonium and nitrate. These comparisons suggest that in spite of the relatively harsh environmental conditions, the N transformations and turnover may be fast in the alpine tundra, as emphasized by several other studies. A number of studies suggest that most of the transformations of N compounds in subalpine and tundra soils may occur under the seasonal snow cover (Freppaz et al., 2007a,b) or in the transition seasons (Brooks et al., 1998; Sickman et al., 2003), resulting in fast production and consumption of N compounds, in a fast turnover of the microbial populations that in turn may lead to a pulse of available C and N compounds in the soil. These fast reactions contribute to the constitution of pools that may be large in size. In contrast to the prolonged isothermal and moist conditions of winter and spring, the summer period is short, variable and probably the hardest period for the microbes to survive (Ley et al., 2004).

Ammonium and nitrate concentrations in topsoils show a clear increasing pattern with decreasing altitude. This pattern is partially interrupted at site IV, where the local topographic conditions (steep slope) may be responsible for a loss of N through severe leaching but also through loss of surface soil horizons by strong erosion processes. Similarly, DON and microbial N concentrations followed the same trend, thus highlighting the increasing capacity of soils to support biological activity and conservation of nutrients with increasing pedogenetic degree, but contemporary suggesting that soil erosion and leaching associated with steep slope act as a limitation for nutrient retention capacity.

At site IV, in fact, the topographic conditions act as a constrain for nutrient retention, and the effect is so pronounced that soil at site IV shows the minimum content of several biologically important compounds. For example, the inorganic nitrogen stock in the top 10 cm at site IV is about 5 kg N ha^{-1}, i.e. lower than reported rates of yearly N deposition at comparable altitudes in the European Alps (cf. Hiltbrunner et al., 2005), while in sites II, III and V IN stock is approximately 9-11 kg ha^{-1}, a value that implies (a) a good retention capacity of the soil with respect to N and (b) a non negligible IN production via biologically-mediated processes, such as organic matter mineralization and nitrification.

These findings highlight the extreme landscape variability that may occur in an alpine tundra environment, with soils of flat areas that may reach a considerable pedogenetic degree, while under steep slopes, pedogenesis may be slowed down because of severe leaching or erosion of part of the soil body. As a result, soils at lower elevation, as at site IV may show a low degree of pedogenesis, comparable to soils at much higher altitude, such as at site I.

Plant Distribution with Respect to Altitude and Nutrient Pools

From the top to bottom positions of the toposequence the synecology of plant populations depicted some clear gradients connected to changes in the most important landscape variables as humidity and nutrients pool: decreasing in bryophyte cover, increasing in vascular plant cover. In the same way, the result of the ordination demonstrated that floristic gradients within the type of plant population coincided significantly with soil properties, thus confirming the results of the former evaluations (Figure 7).

Soils at higher plots (I and II) are located in flat areas and are therefore protected from erosion and accumulation of materials. At these sites, however, the harsh environmental conditions and the short time lag after the disappearance of perennial snow, combined with the slow and incomplete organic matter decomposition, which was also suggested in the Swiss Alps (Hiller et al., 2005), result in poorly developed soils, colonized by pioneer species. This hypothesis is also corroborated by the presence of some bryophytes such as *Polytrichum sexangulare* and *Kiarea falcata* and the epatic *Anthelia juratzkana*.

Consistently with soil chemical properties previously discussed, demanding species become more and more important with decreasing elevation, a pattern also confirmed by the ordination. At higher sites, we found less demanding species such as *Poa laxa*, *Gnaphalium supinum* and *Salix erbacea*, while at lower elevation more demanding taxa were described, with presence of *Poa alpina*, *Festuca violacea*, *Cirsium spinosissimum* and *Viola bifora*. The richer nutrient pool at lower sites also determines the increase in number of species, less and less affected by pioneering conditions. The same vegetation pattern may be related to soil pH increase with decreasing altitude. In such a context, the extreme borders are represented by *Cardamine resedifolia* (sites I and II, lower pH) and *Galium megalospermum* (site IV, higher pH). This increase in pH may also reflect changes in geology, with gneiss and micaschists at sites I and II, calcschists at site IV and ophiolites at site V. Moreover, the light availability, controlled by morphology, seems to play an important role in plant composition. The occurrence of north-facing, steep slope becomes increasingly important with decreasing altitude. As a result, sites I and II, although at higher elevation, benefit from a higher amount of light, that also increases the daily thermal sums. Consistently, at higher sites sun species such as *Minuartia sedoides*, *Saxifraga bryoides* and *Cerastium uniflorum* were dominant, while at sites IV and V the appearance of more tolerant species (*Homogyne alpina*, *Vaccinium myrtillus* and *Viola bifora*) was recorded.

In terms of subplot variability, as shown in figure 7, sites at higher elevation show the highest subplot-to-subplot variability and heterogeneity, while, once again, at lower elevation this variability decreases. These findings are consistent with the presence of pioneer plants at sites I and II, that may also be due to a more recent (i.e., on the order of a century) disappearance of glacier ice sheets. This implies a strong microvariability that may lead to substantial differences at the small scale, mainly related to the timing and the distribution of the snow cover (Hiller et al., 2005).

Richard (1973) reported that the shift from pioneer to stable vegetation cover needs stable and homogeneous substrata, the formation of whose would require more than 200 years at an elevation of 2000 m asl.

CONCLUSION

Soils and plants in the Alpine Tundra environment are a mosaic of different landscapes strongly controlled by topographic conditions that in turn control water and snow cover distribution. These factors drive to very different pedo-environments even at the small scale.Despite this, we found some clear trends that could be associated with the altitude. The plan t species abundance and the biodiversity, the pedogenesis and the nutrient retention capacity display an overall increasing trend with decreasing altitude. Given that, topography plays a major role in determining the pedogenetic degree and the retention capacity of soil of the Alpine Tundra belt. Soils under steep slopes such as at site IV may in fact show properties comparable with soils located several hundred meters above along an elevation gradient.

The relatively large nutrient pools of these alpine Tundra soils suggest that biological processes are active in spite of the harsh environmental conditions. Given that the bulk of studies conducted in snow covered ecosystems suggest that much of biological processes occur during winter, under-snow biology may be the main responsible for significant transformations of C and N pools.

Plant species distribution and characteristics overall reflect the edafic status of soils. More developed soils support a plant community that is larger in size and characterized by a higher biodiversity. The effect of topography, so strong for the soil development, also affected the plant community. Moreover, plant species distribution reveals other microtopography-related controls, i.e. the occurrence of more sun species at flat areas. In this sense, plants appear to be more sensitive than underlying soils to small changes in factors that control their development. As shown by Darmody et al. (2004) altitude and latitude control the general climate, but microclimate is largely controlled by topoclimatic factors, snow cover, soil moisture, exposure, aspect, etc. Snow cover has a great role in modifying the microclimate, and the soil moisture is controlled by soil texture as influenced by microtopography.

Given the vulnerability of alpine tundra soils to and the attractiveness of alpine zones as destinations, the potential for negative recreation impacts will only continue to increase. And if more areas of alpine tundra are eroded to bare bedrock, then there is great risk that the unique plants of the alpine zone will be extirpated from this region (Munroe et al., 2007). It is imperative, therefore, to develop management strategies built upon a solid foundation of field-based science to ensure the preservation of these specific pedoenvironments.

Acknowledgments

This study was funded by the Monterosa 2000 S.p.A. in the framework of the "Convenzione per il monitoraggio e recupero ambientale inerenti il progetto di impianto di innevamento programmato sulla pista di sci Olen e opere connesse". Thanks to the Comando Truppe Alpine-Servizio Meteomont for the nivo-meteorological data.

REFERENCES

Bockheim, J. G., Munroe, J. S., Douglass, D. & Koerner D. (2000). Soil development along an elevational gradient in the southeastern Uinta Mountains, Utah, USA. *Catena, 39*, 169-185.

Bowman, W. D., Theodose, T. A. & Schardt J. C. (1993). Constraints of nutrient availability on primary production in two alpine tundra communities. *Ecology, 74*, 2085-2097.

Brookes, P. C., Landman, A., Pruden, G. & Jenkinson D. S. (1985). Chloroform fumigation and the release of soil nitrogen. A rapid direct extraction method to measure microbial biomass nitrogen in soil. *Soil Biology & Biochemistry, 17*, 837-842.

Brooks, P. D., Williams, M. W. & Schmidt, S. K. (1996). Microbial activity under alpine snowpacks, Niwot Ridge, Colorado. *Biogeochemistry, 32*, 93-113.

Brooks, P. D., Williams, M. W. & Schmidt, S. K. (1998). Inorganic nitrogen and microbial biomass dynamics before and during spring snowmelt. *Biogeochemistry, 43*, 1-15.

Brooks, P. D. & Williams, M. W. (1999). Snowpack controls on nitrogen cycling and export in seasonally snow-covered catchments. *Hydrological Processes, Vol. 13*, 2177-2190.

Burns, S. F. & Tonkin, P. J. (1982). Soil-geomorphic models and the spatial distribution and development of alpine soils. In: C. E. Thorn (Editor), Space and Time in Geomorphology. *George Allen and Unwin*, London, 25-43.

Butler, D. R. & Malanson, G. P. (1999). Site locations and characteristics of miniature patterned ground, eastern Glacier National Park, Montana, USA. *Landform Analysis, 2*, 45-49.

Cortini Pedrotti C. (2001-**2006**). **Flora dei Muschi d'Italia**. A. Delfino editore, Roma 2 voll.

Crooke, W. M. & Simpson W.E. (1971). Determination of NH_4^+ in Kjeldahl Digests of Crops by an Automated procedure. *Journal of the Science of food and agriculture, 22 (1)*, 9-10.

Darmody, R. G., Thorn, C. E., Schlyter, P. & Dixon, J. C. (2004). Relationship of Vegetation Distribution to Soil Properties in Kärkevagge, Swedish Lapland. Artic, *Antartic and Alpine Research, 36 (1)*, 21-32.

Dirnböck, T., Dullinger, S. & Grabherr, G. (2003). A regional impact assessment of climate and land-use change on alpine vegetation. *Journal of Biogeography, 30*, 401-417.

Edwards A.C., Scalenghe R., Freppaz M. (2007) Changes in the seasonal snow cover of alpine regions and its effect on soil processes: a review. *Quaternary International* 162-163: 172-181.

Egli, M., Mirabella, A. & Fitze, P. (2003). Formation rates of smectites derived from two Holocene chronosequences in the Swiss Alps. *Geoderma, 117*, 81-98.

Egli, M., Mirabella, A., Sartori, G. (2004). Weathering of soils in alpine areas as influenced by climate and parent material. *Clays and Clay Minerals, 52*, 287-303.

Egli, M., Sartori, G., Mirabella, A., Favili, F., Giaccai, D. & Delbos, E. (2009). Effect of north and south exposure on organic matter in high Alpine soils. *Geoderma, 149*, 124-136.

Filippa, G., Freppaz, M., Williams, M. W., Liptzin, D., Seok, B., Chowanski, K., Hall, B. & Helmig, D. (2009). Winter and summer nitrous oxide and nitrogen oxides fluxes from a seasonally snow-covered subalpine meadow at Niwot Ridge, Colorado. *Biogeochemistry* vol. 95 (1): 131-149.

Fisk, M. C. & Schmidt, S. K. (1995). Nitrogen mineralization and microbial biomass N dynamics in three alpine tundra communities. *Soil Science Society America Journal, 59*, 1036-1043.

Freppaz, M., Williams B. L., Edwards, A. C., Scalenghe, R. & Zanini, E. (2007a). Labile nitrogen, carbon, and phosphorus pools and nitrogen mineralization and immobilization rates at low temperatures in seasonally snow-covered soils. *Biology and Fertility of Soils, 43*, 519-529

Freppaz, M., Williams B. L., Edwards, A. C., Scalenghe, R. & Zanini, E. (2007b). Simulating soil freeze/thaw cycles typical of winter alpine conditions: implications for N and P availability. *Applied Soil Ecology* 35: 247-255.

Freppaz, M., Maggioni, M., Gandino, S. & Zanini, E. (2008a). Snowpack evolution on permafrost, non-permafrost soils, and glaciers in the Monte Rosa Massif (Northwest Alps, Italy) Poster Session. 9th International Conference on permafrost, Fairbanks-Alaska, 29th June – 3rd July 2008. D. L. Kane, & K. M. Hinkel (Eds.), 79-80.

Freppaz, M., Marchelli, M., Celi, L. & Zanini, E. (2008b). Snow removal and its influence on temperature and N dynamics in alpine soils (Vallée d'Aoste - NW Italy). *Journal of Plant Nutrition and Soil Science, 171*, 672-680.

Glaser, B., Turrión, M. B., Solomon, D., Ni, A. & Zech, W. (2000). Soil organic matter pools in mountain soils of the Alay Range, Kyrgyzia, affected by deforestation. *Biology and Fertilty of Soils, 31*, 407-413.

Hiller, H., Nuebelt, N., Brollt, G. & Holtmeiert, F. (2005). Snowbeds on Silicate Rocks in the Upper Engadine (Central Alps, Switzerland) - Pedogenesis and

Interactions among Soil, Vegetation, and Snow Cover. *Arctic, Antarctic, and Alpine Research. 37(4)*, 465-476

Hiltbrunner, E., Schwikowski, M. & Körner, C. (2005). Inorganic nitrogen storage in alpine snowpack in the Central Alps (Switzerland). *Atmospheric Environment, 39*, 2249-2259.

Hitz, C., Egli, M. & Fitze, P. (2002). Determination of the sampling volume for representative analysis of alpine soils. *Zeitschrift fur Pflanzenernahrung und Bodenkunde, 165*, 326-331.

IUSS Working Group WRB. (2006). World Reference Base for Soil Resources 2006. World Soil Resources Reports n. 103. FAO, Rome.

Killham, K. (1994). Soil Ecology. Cambridge University Press, Cambridge.

Klimowicz, Z. & Uziak, S. (1996). Arctic soil properties associated with micro-relief forms in the Bellsund region (Spitsbergen). *Catena, 28*, 135-149.

Körner, Ch., Diemer, M., Schappi, B. & Zimmermann, L. (1996). Responses of alpine vegetation to elevated CO_2. In: Koch GW, Mooney HA (eds.) *Carbon dioxide and terrestrial ecosystems*. Academic Press, San Diego, 177-196.

Körner, C. (2003). Alpine Plant Life. Functional Plant Ecology of High Mountain Ecosystems. *Springer-Verlag Berlin Heidelberg*, 343.

Kruskal, J. B. (1964). Nonmetric multidimensional scaling: a numerical method. *Psychometrica, 29*, 115-129.

Landolt, E. (1977). Ekologische Zcigerwerte zur Schweizer Flora. Veroffentlichungen des Geobotanischen Instituts der ETH, Stiftung Riibel. Zurich, *64*, 208.

Ley, R.E., Williams, M.W. & Schmidt, S.K. (2004). Microbial population dynamics in an extreme environment: controlling factors in talus soils at 3750 m in the Colorado Rocky Mountains. *Biogeochemistry, 68*, 313-335.

Macdonald, B. C. T., Melville, M. D. & White, I. (1999). The distribution of soluble cations within chenopod-patterned ground, arid western New South Wales, Australia. *Catena, 37*, 89-105.

Makarov, M. I., Volkov, A. V., Malysheva, T. I. & Onipchenko V. G. (2001). Phosphorus, nitrogen and carbon in the soils of subalpine and alpine altitudinal belts of the Teberda Nature Reserve. *Eurasian Soil Sci., 34*, 62-71.

Makarov, M. I., Glaser, B., Zech, W., Malysheva, T. I., Bulatnikova, I.V. & Volkov, A. V. (2003). Nitrogen dynamics in alpine ecosystems of the northern Caucasus. *Plant and soil, 256*, 389-402.

Mather, P. M. (1976). Computational methods of multivariate analysis in physical geography. J. Wiley & Sons. London.

McCune, B.& Mefford, M. J. (1999). PC-ORD. Multivariate analysis of ecological data, version 4.34 – MjM Software Design. *Gleneden Beach*, Oregon, U.S.A.

Mirabella, A., Egli, M., Carnicelli, S. & Sartori, G. (2002). Influence of parent material on clay minerals formation in podzols of Trentino - Italy. *Clay Minerals*, 37, 699-707.

Mirabella, A. & Sartori, G., (1998). The effect of climate on the mineralogical properties of soils from the Val Genova Valley (Trentino, Italy). *Fresenius Environmental Bulletin*, 7, 478-483.

Munroe, J. S., Farrugia, G. & Ryan, P. C. (2007). Parent material and chemical weathering in alpine soils on Mt. Mansfield, Vermont, USA. *Catena*, 70, 39-48.

Nadelhoffer, K. J., Giblin, A. E., Shaver, G. R. & Linkins, A. E. (1992). Microbial processes and plant nutrient availability in arctic soils. In: Arctic Ecosystems in a Changing Climate. An ecophysiological perspective. Eds. F S Chapin III, R L Jefferies, J F Reynolds G R Shaver and J Svoboda. pp. 281–300. Academic Press, San Diego, CA.

Neff, J. C., Bowman, W. D., Holland, E. A., Fisk, M. C. & Schmidt S. K. (1994). Fluxes of nitrous oxide and methane from nitrogen-amended soils in a Colorado alpine ecosystem. *Biogeochemistry*, 27, 23-33.

Peterson, E. B. & McCune, B. (2001), Diversity and succession of epiphytic macrolichen communities in low-elevation managed conifer forests in Western Oregon. *Journal of Vegetation Science*, 12, 511-524.

Post, W. M., Emanuel, W. R., Zinke, P. J. & Stangenberger, A. G. (1982). Soil carbon pools and world life zones. *Nature*, 298, 156-159.

Richard, L. (1973) Dynamique de la végétation au bord du grand glacier d'Aletsch (Alpes suisses), *Berichte der Schweizerischen botanischen Gesellschaft.* 83, 159-174.

Schumacker, R. & Vaňa, J. (2000). Identification Keys to the Liverworts and Hornworts of Europe and Macaronesia (Distribution and status)1st edition. Documents de la Station scientifique des Hautes-Fagnes n° 31

Sickman, J.O., Leydecker, A., Chang, C.C.Y., Kendall, C., Melack, J.M., Lucero D.M. & Schimel, J. (2003). Mechanisms underlying export of N from high-elevation catchments during seasonal transitions. *Biogeochemistry*, 64, 1-24.

SISS – Italian Soil Science Society (1998). Metodi di analisi fisica del suolo. Franco Angeli Editore, Milano.

SISS – Italian Soil Science Society (2000). Metodi di analisi chimica del suolo. Franco Angeli Editore, Milano.

Soil Survey Division Staff. (1993). Soil survey manual. Soil Conservation Service. U.S. Department of Agriculture Handbook 18.

Stanton, M. L., Rejmanek, M. & Galen C. (1994). Changes in vegetation and soil fertility along a predictable snowmelt gradient in the Mosquito Range, Colorado, USA. *Arctic and Alpine Research 26*, 364-374.

Tutin, T. G., Heywood, V. H., Burges, N. A. & Valentine, D. H. (1993). Flora Europaea Cambridge University Press. Cambridge.

Walker, B. & Steffen, W. (1997). An overview of the implications of global change for natural and managed terrestrial ecosystems. *Conservation Ecology, 1*, 2.

Williams, M. W., Losleben, M. V. & Hamann, H. B. (2002). Alpine areas in the Colorado Front Range as monitors of climate change and ecosystem response. *Geogr. Rev., 92*, 180-191.

Wu, G., Wei, J., Deng, H. & Zhao J. (2006). Nutrient cycling in an Alpine tundra ecosystem on Changbai Mountain, Northeast China. *Applied Soil Ecology, 32*, 199-209.

Zhang, T., Barry, R. G., Knowles, K., Heginbottom, J. A., Brown, J. (1999). Statistics and characteristics of permafrost and ground-ice distribution in the northern hemisphere. *Polar. Geog., 23*, 147-169.

In: Tundras: Vegetation, Wildlife...
Editors: B. Gutierrez et al. pp. 111-149
ISBN: 978-1-60876-588-1
© 2010 Nova Science Publishers, Inc.

Chapter 4

MODERN CLIMATE TRENDS AND POSSIBLE CHANGING OF ARCTIC COASTAL ZONE (RUSSIAN SECTOR)

[1]Sergey Nikiforov[*], [2]Vladimir Byshev[**], [3]Ogorodov Stanislav[***] and [4]Putans Victoria[****]

[1]Doctor of Sciences, main scientific researcher,
Shirshov Institute of oceanology.
[2]Doctor of Scienses, Head of Lab, Shirshov Institute of oceanology.
[3]Leader researcher, Lomonosov Moscow State University,
Department of Geography.
[4]Shirshov Institute of oceanology.

ABSTRACT

The current climate variability allows establishing that natural processes are in line with the most important and observed global warming factors. The climate system in mid 70-ties of the previous Century entered an equilibrium which brought about a rapid global reallocation of the atmosphere. The concomitant alteration of the planetary atmosphere circulation brought about

---

[*] Corresponding authors: E-mail: nikiforov@ocean.ru
[**] E-mail: labbyshev@ocean.ru
[***] E-mail: ogorodov@aha.ru
[****] E-mail: vitapu@mail.ru

intensification of meridional warmth transfer from the Indian Ocean to the central areas of the Eurasian Continent and from the Pacific Ocean to Alaska. It was exactly in those areas that the large scale subsurface temperature anomalies in the last quarter of the XX-th century took place. The first decade of the XXI century revealed the features characteristic of the climatic system return to the state antecedent to the mid 70-ties of the previous century. Thermodynamic conditions of Arctic Ocean are defined by a number of factors, among which of greatest importance are:

- basin waters circulation and exchange with Atlantic and Pacific Oceans;
- heat flow balance on the Arctic basin surface and presence of energetically active zones with intensive ocean-atmosphere interaction and transformation of Atlantic waters;
- regional atmospheric circulation and connection thereof with planetary circulation.

Colligation of hydrogeological data provided by various Arctic expeditions in past century leads to the deduction of considerable warming (for 0.5-1.5C) of Atlantic waters in Arctic basin observed in last half of XX Century. Arctic warming inevitably affects all components of natural environment, including dynamic and morphology processes in coastal zone of Arctic seas. Simultaneously with climate warming the rising of sea level occurs. This also is a very important factor of future changes in coastal zone.

Decrease of area and existing time of ice sheet in Arctic Ocean seas and, as a consequence, increase of hydrodynamic activity in coastal zone will undoubtedly be one of the most important factors of changes for coastal zone dynamics in XXI century.

The most substantial changes in coastal zone dynamics and morphology will occur in Arctic seas with unconsolidated permafrost environments and intensive thermal erosion. The velocity of shores distraction is evidently been largely determined by deposits composition and iciness, both dependent on lithogenic environment and following cryogenic reconstitutions. Modern velocities of Russian Arctic shores thermal erosion vary widely from first meters to tens and more over per year in general. Rapid changes in environment can tilt a fragile balance and cause shore degradation. This is particularly important for Arctic shores which erode and step back nearly everywhere. Global warming aggravates the situation. Ice-free period extension contributes to increase in total near-shore wave energy. Ice-verge back off in summertime brings about the same effect of wind acceleration enlarging. The mentioned changes can have different influence on different shores. Thermal erosion will definitely increase. Thermal erosion, thermal denudation, deflation can bring about considerable economic loss and bring to naught profitability of raw material extraction.

INTRODUCTION

The climate system in mid 70-ties of the previous century entered an equilibrium state. The concomitant alteration of the planetary atmosphere circulation had intensified meridional warmth transfer from Indian Ocean to central areas of Eurasian Continent and from Pacific to Alaska. It was exactly in those areas that the large scale subsurface temperature anomalies in the last quarter of the XX-th century took place. The first decade of the XXI century revealed typical signs that the climatic system has started to return to the state antecedent to mid 70-ties of the previous century.

Cyclonic and anticyclonic activity in atmosphere and ocean determines global conditions and weather variability [Monin, 1969]. **Cyclogenesis and** anticyclogenesis are associated with baroclinic unstability mechanism which performs large-scale turbulent exchange of masses and energy. This exchange is directed to destruction of permanently forming vertical and horizontal global stratification. Energy and cinematic of vortex disturbances are observed by global meteorological network for more than a century, so fluctuations in cyclogenesis intensity and changes of general transfer traces of whirls have been detected [Byshev, 2003].

The most significant climate variations appear when turbulent transfer in atmosphere changes its direction from zonal to meridional and otherwise. Each transfer type domination period is characterized by different climatic scenarios. Recent weather anomalies are indicated nearly elsewhere, and are most certainly indicators of climate change. Whether these changes are natural or antropogenic, is important for estimation of long-period tendencies in global and regional weather formation.

Current weather pattern have been forming as superposition of different global and regional features of modern climate [Hasselmann, 1976; Hasselmann, Frankignoul, 1977]. These features indicate random behavior of climate system; variations could be predicted only in probability. In the same time ocean and atmosphere interaction is proved by statistically valid signs, which are deducted from simultaneous analysis of long-time measurement runs [Chiang, Kushnir, 2000; Latif, 2001; Mo, Hakkinen, 2001]. The main aim of climate system investigation is to predict probable climate scenarios, their evolution and establishment terms. Climate changes sources and formation mechanisms of large-scale thermodynamical anomalies are supposed to be known for any predictions. The most significant climate changes are observed in mid- and high latitudes as well as in Arctic.

Modern climate warming will inevitably cause the rising of world ocean level and changes in oceanologocal, geomorphological, geological, biological, and other parameters of natural environment at coastal zone and on shelf. But changes mentioned will be irregular. The largest are expected in Arctic.

Thermodynamic regime of Arctic Ocean is defined by several factors. The main are: (1) the basin water circulation itself and water exchange with Atlantic and Pacific oceans; (2) heat-flow balance on surface and presence of active areas with intensive atmosphere - Atlantic water interaction; (3) regional atmospheric circulation and its relation to the global one.

Atlantic water flow into Norway Sea southward Medvejiy Island (70 N). The east branch (Nordcap Current) spreads in Barents Sea as surface water mass. After cooling and freshening, its density increase and the water down lifted on medium depth. Having passed Barents Sea, the current enters northern part of Kara Sea and afterwards travel to Nansen Basin through St.Ann Channel.

The west branch carries Atlantic waters northward along Svalbard continental slope. Part of this flow going south in Fram Strait together with East-Greenland water on medium depth. The rest, as boundary current in medium layer, travels north through Fram Strait into Arctic basin, then turns east, go along continental slope of Nansen basin and enters north parts of adjacent seas through deep channels. While their travel, Atlantic waters loose warmth, become denser, and sink.

Colligation of Arctic expeditions hydgrological data over last 100 years result in conclusion that Atlantic waters in Arctic basin had had a considerable warming (for 0.5-1.5) in second half of XX Century. At the same time the anomal core of warm water occurs in upper part of medium layer (150-250m). This fact shows an increase in thermochaline structure vertical stability and weakening in convective mixing mechanism.

Variations in the field of ground air temperature and temperature of ocean surface in north hemisphere were investigated for a long time. These investigations had shown that secular **tendencies were in "phase opposition" over** oceans and continents, as well as over the east and the west parts of Arctic [Byshev, Neyman, Romanov, 2005; 2006].

In first half of XX Century while anomalies of surface temperature in Atlantic and Pacific oceans had been in a high-rate rise, the reduction (opposite second derivative signs) was observed over the continents (Siberian, American, European and Far Eastern sectors).

In second half of XX Century the situation reversed: temperature was in a rise over continents and in a slow-down over oceans. Therefore large-scale secular redistribution of warmth between oceans and continents was observed in global system ocean-atmosphere-continent. Structural changes in athmospheric circilation over the Northern hemisphere were shown to be related to peculiarities of global variability of near-ground temperature field. On this base the warming of weast Arctic could be concerned as a consequence of structural athmospheric circulation reorganization [Byshev, Kononova, Neyman, Romanov, 2002; 2004].

We had been investigating variability of contemporary Northern hemisphere climate by well-known database on global anomaly fields of near-ground temperature [Jones, Moberg, 2003] and atmospheric pressure. Global warming had attracted attention at mid-70s. The relatively fast reorganization had occurred in global atmospheric pressure field at that period [Byshev, Neyman, Romanov, 2009]. To get non-random (statistically valid) quantitative estimation of reorganozation mentioned, we residuated mean values of global climatic fields of sea-level atmospheric pressure and near-surface temperature for two 25years-long periods: (1975-1999) and (1950-1974). The same method was used for monthly residuations in question. All these calculations allow retracing temperature and pressure fields climatic transformation from season to season.

The atmospheric pressure field rebuilding have resulted in global atmosphere circulation change. Quantitative estimation of this change was obtained from acquainted semigeostrophic correlation, which gives acceptable description of global wind velocity field in lower throposphere but for relatively tight equatorial zone (Rossby deformation radius) [Gill, 1986]. Relatively fast reorganization of global atmospheric pressure fields was followed by large-scale positive and negative pressure anomalies (Figure.1). For example, sea-level atmospheric pressure had had a considerable low over Polar Regions (Arctic and Antarctic) and over the most part of Pacific Ocean. In Northern hemisphere pressure anomalies were mainly negative and had minimums in

(1) northwest Canada and Greenland, including Baffin Sea
(2) Ob and Yenisei mouths and south part of Kara Sea
(3) Bering Sea.

At the same time atmospheric pressure had a rise over some regions. There were bents of high pressure in both hemosphere midlatitudes and local maximums over

(1) central Atlantic and Midterranian seas

(2) central China
(3) Antarctic Circumpolar Current southward Africa
(4) Australia
(5) Drake Strait and South America tail.

Features of global deformations in question (occurred in mid 70ties) were accompanied by certain changes in general atmospheric circulation. Arrows on Figure.1 show wind anomalies field vectors. East peripheries of low pressure areas, mentioned above, have shown an intencification of warm air masses transfer from low to high latitudes, especially over north-western North America and central Siberia. The first area had rise in temperature of near-surface air because of mid-Pacific tropical masses inflix, the second – because of intensification in meridional fluxes from Indian Ocean.

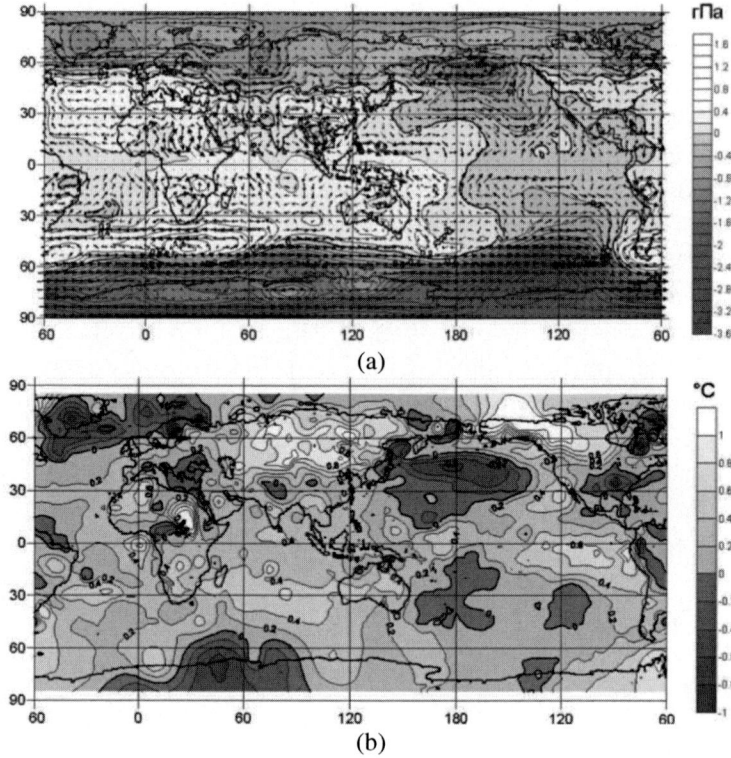

(a) global atmospheric pressure field on sea level
(b) near-surface temperature

Figure. 1. Climatic changes of in XX Century last quarter (1975-1999), arrows show geostrophic wind velocity.

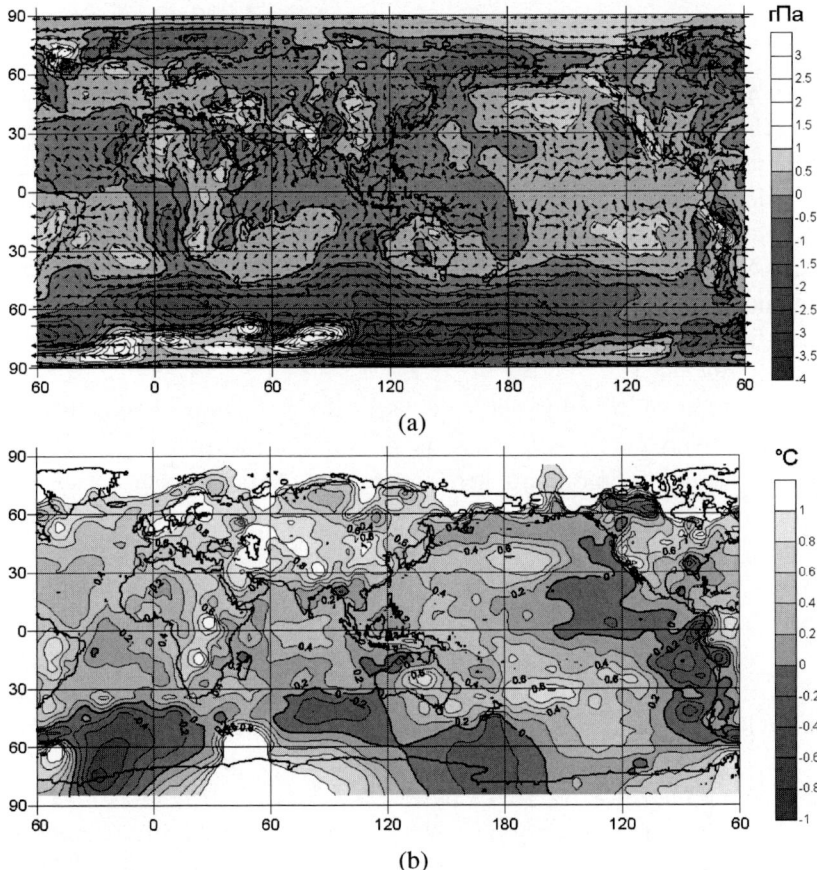

(a) global atmospheric pressure field on sea level
(b) near-surface temperature

Figure.2. Climatic changes in 2000-2008.

In contrast, west peripheries had intensification of cold air masses from high latitudes to low ones. Here in last quater of XX Century took place outfluxes of Arctic air to Pacific and Atlantic, and to Eurasian continent as well (westward 60° longitude). It ought to be remarked, that these outfluxes are responsible for heightened contrasts between near-ground temperature anomalies (Figure 1b). Arctic air entering warm Atlantic and Pacific surface contributed to intensification of heat exchange between ocean and atmosphere. This effect is fortified by large-scale negative temperature anomalies in ocean subsurface layer.

Areas with anomaly high atmospheric pressure are characterized by meridional air mass transfer: from high latitudes to low ones on west peripheries, and inverse on east peripheries. For example, negative near-surface temperature anomaly in east Mediterranean and Black Sea obviously belongs to east periphery of high-pressure anomaly and its intensive border transfer of cold arctic air. The same mechanism is the base for vast positive near-surface anomaly in the central area of Euroasian continent, and negative anomaly in Japan Sea and in Pacific southward and south-eastward Japan Isl.

In south hemisphere interrelation between atmospheric pressure anomalies and near-surface temperature is the same. The most representative example is high-latitude area of South America. Westward Drake Straite warm air masses are tranfered from low latitudes to high ones, eastward – cold masses travel form high lalitudes to low ones.

Investigation of monthly climatic variability of atmospheric pressure and near-surface temperature anomalies is of certain interest. In cold year-time (November-March for northern hemisphere and April-October for South one) negative anomalies are deepen, because cyclonic turbulence of atmospheric circulation is intensified. On the contrary, in warm year-time (April-June and December-March accordingly) cyclonic turbulence decrees and interlatitudal heat exchange falls. Estimation of climatic features of annual near-surface temperature variations has shown that in northern hemisphere anomalies have higher seasonal activities. From December to May near-surface temperature in North Atlantic and North Pacific had a regular fall to extreme on -0.8. After that, the anomaly remained negative but absolute value was decreasing, possibly because of earth surface radiation heating.

Such physical interpretation of climatic processes variability is confirmed by special observations of hydrophysical parameters. The observations have been obtained in different ocean experiments during 1975-1999, for example the largest world-wide oceanographic projects Megapolygon and Atlantechs-90 ["MAGAPOLYGON experiment", 1992; Byshev, Koprova, Navrotzskaya et al., 1993].

The first one had occurred from July to November of 1987 in Pacific Ocean eastward Japan islands, where negative climatic anomaly of surface temperature takes place (Figure.1b). Estimations of heat flows over the ocean surface showed negative warmth balance in period from August to October. This implies that evaporation heat output, turbulent heat flow from ocean to atmosphere, and ocean surface long-wave radiation had exceeded sun radiation heat input. Annual variation amplitude of ocean surface temperature in 1986-1987 had exceeded its average climatic value. To our opinion this fact is due to convective cooling of significantly thick upper active layer while summer warming in thin subsurface layer [Byshev, Neyman, Romanov, 2006].

The well-known Atlantic experiment was carried out in Newfoundland area in April-June, 1990. Data analysis showed the large-scale negative anomaly. It was -2.0-3.0 on surface and had involved at least 1000m downward. In 1975-1999 formation of negative temperature anomalies on ocean surface was common in east part of Mediterranean Sea as well [Zveryaev, Arhipkin, 2008].

Thus, in last quarter of XX Century negative anomalies of near-surface temperatures were formed in Atlantic and Pacific oceans and in Mediterranean Sea (Figure.1b). This situation is caused by Arctic masses outflows, which occur in order of atmospheric pressure field distribution features (Figure.1a). As is shown above, such conclusion is supported by independent data.

Comparison of average hydrometeorological fields characteristics in 1975-1999 and 2000-2008 periods is of particular interest. Pressure extremes are observed to have had sign reversal nearly everywhere. Areas with previous (1975-1999) negative anomalies of atmospheric pressure have been showing tendencies to pressure increase, and vice versa (decreasing tendencies in positive anomalies zones). The near-surface temperature field has shown opposite tendencies (Figure.1b and 2b). A next phase change in interdecade variation of global climatic system could be derived from these facts. The change is caused by internal dynamic of the system.

So, near-surface temperature dynamic is time-irregular. Nevertheless the last tendency of temperature increasing had led to decrease in Arctic drift-ice areas which in turn caused storm factor activation, increase in seasonally thawed layer thickness, and speed up shore retreating. Dynamic of ice-covered area could be well observed on Arctic and Antarctic Institute ice maps (www.aari.bw.ru). These maps were obtained from satellite photos and could be used for quite reliable comparison analysis of ice conditions since 1972. The ice-covered area considerably changes through seasons and years. The comparative analysis requires division into years with minimal and maximal ice cover.

Minimal areas fixed by southern border of drift ice in September that occur northward Severnaya Zemlya and New Siberian Islands. It means that the wide bent near Chaun Bay is ice-free, and Vilkitski, Dmitriy Laptev, Long and Sannikov Straits are open also. In this case, the clearing northward Bering Strait (in Chukchy Sea) appears in May while in June whole south-east zone is ice-free; in period from July to October all sea but for north margin is open.

If ice cover is maximal, the border of drift ice in September comes against shores of Laptev and East Siberian seas. Vilkitskiy, Dmitriy Laptev, Long and Sannikov Straits are tightly packed with ice; the only open area is south-east Chukchy Sea.

Comparison shows: during last decade changing in ice situation have been leading to establishment minimal ice cover as a common one (table 1).

Quternary history shows wide World ocean level changes. There are: "glacial" and "interglacial" levels, which have changed in 10-thousand year cycles and had amplitudes of tens to hundreds meters (first order periodicity); thousand-year variations (second order periodicity); hundred-year variations (third order periodicity). Of course there are variations with shorter periods.

According to models of past climates, the global warming should result in Arctic environments similar to climatic optimum of Riss-Wurm interglacial period. In that time ice sheet is believed to be only on Greenland, while Arctic ice sheet had melted because of eastward warm waters intrusion.

Future Ocean's level rise in next 100 years are varied from 10-20sm to 4m. The main reason for such a difference lies in complexity of any forecasts, especially of Greenland and Antarctic ice sheets reaction. Annual melting of continental and gletcher ice leads to incresed water outflow and consequently world ocean level rise on 1.5-2mm/year in average (up to 2.6mm/year in Arctic Ocean) (Klige, 1998 and others). Such slow level rising should not cause rapid and intensive shore destruction; wind setups should be of more importance due to their amplification, caused by increase in wave acceleration distance.

Table 1. Ice cover changing within 1972 – 2005.

Years	Maximum ice cover	Minimum ice cover
1972-1981	4	2
1982-1991	3	2
1992-2005	1	10

At the same time every Arctic sea has its own peculiarities, both in past, modern and future development. Norwegian and Barents seas have got minimal changes in shore and shelf zones in XXI Century. Fundamental changes are supposed to occur in hydrological pattern. The warm Atlantic water inflow to Barents Sea will increase. Due to this, drift ice will be retreating to north-west during spring-summer time. Consequently, all offshore area west of Novaya Zemlya will be spare of any ice, even seasonal.

In winter drift ice border will exist on line from north end of Novaya Zemlya to south end of Svalbard. Relatively warm Atlantic water will intensively pass to Kara Sea through Sedov Channel. Melting of Novaya Zemlya glaciers will cause increase in terrigenic material outflow. In second half of XX Century the total glacial material output from Novaya Zemlya was near 50mln tons/year. We suppose increasing up to 70-75 mln.ton/year. Glacial material will rapidly deposit offshore Novaya Zemlya.

In Kara Sea average annual temperature and Atlantic masses inflow will increase, hence water warming is predicted. By the end of XXI Century west part of Kara Sea would be ice-free for at least 4 months per year (July-October), while in XX Century this period was 2.5-3 months. This process may spread over east part of the basin, so that in summer and autumn there will be favorable navigation conditions offshore Kara Sea. In turn, prolonged clear water conditions will cause intensification of wave regime, which will most obviously affect dynamics of thermoabrasion and accumulation coasts (west Yamal, etc.).

In Laptev Sea storm frequency and ice-free time will increase, the last for 50% (from 2 months to 3, on our opinion). Consequently, wave action on coastline and bottom in XXI century will be intensified nearly in 1.5-2 times. This will deepen thermal abrasion of coasts so sediment supply (organic material as well) also will increase.

In East-Siberian Sea climate warming will cause the most significant environment changes, mainly in ice-cover (decrease), thermal abrasion (acceleration) and change of sediment regime in coastal zone and on the shelf.

Accumulation coasts development of Arctic seas, Chukchy Sea in particular, will be accompanied by transgression. Cathastrophic changes and erosion of coastal relief are not to wait here.

Coastal zone changes mostly in permafrost environments. Defrosting, soil creeps to cliff base, where even slight waves wash it out. Such environment is characterized by particular slight coast line; this can be seen in the west part of East-Siberian Sea. Predicted climate changes should raise thermal abrasion up to 20-40 m/year. The same changes would occur in Laptev and Kara Seas.

Opened in early XIX Century, some not large (appr. 10-20km long) islands had gone by now because of thermo abrasion effect. For example, in Laptev Sea the Vasilevskiy Island had lessen from 7km in length and 0.5km in width in 1823 to 4.6km length in 1912, and died away by 1936. The Semenovskiy island was 14km long in 1923, 2km long and 0.5km width in 1936, 1620m long and 236m width in 1945, and died out by 1951 [Belov, 1966, Are, 1980].

It should be noted that in Laptev Sea since 2004 and especially in 2007 an intensification of coastal processes is observed. These are slope processes: solifluction (cryosolifluction) first of all, and rapid increase in erosion of shores with thermal abrasion and thermal denudation. Within some monitoring areas have shown retreating velocities 1.5-2.5 times more than average secular ones. At the present time the fastest shore erosion is observed in glacial areas. Maximum average annual retreating rate exists on north point of Moustah island in Laptev

Sea (13m/year) and westward Krestovskiy cape in East-Siberian sea (12m/year) [Grigoriev, 2008].

The thermal influence shows itself as energy transmission to the coast, which is composed of frozen sediments, via radiative and sensible heat fluxes from air and water. Accordingly, higher air and water temperatures, together with a longer ice-free period and longer period with positive air temperature, affect the stability of frozen coasts. The role of the thermal factor usually increases with increasing ice content in coastal deposits. In turn, low ice content renders the wave-energy factor more significant.

Coastal bluffs formed of frozen deposits with low ice content are not subject to thaw slumping, permafrost creep (solifluction), gully thermoerosion and thermokarst. Periodicity of extreme storm surges and the total wave energy activity in the coastal zone during the active dynamic period are the main factors which determine the dynamics of coasts with low ice content. We could suppose that in the case of climate warming, low ice content coastal dynamics will have similar features to those which they have in the warmest years and decades at the present.

So, rising temperatures are altering the arctic coastline and much lager changes are projected to occur during this century as a result of reduced sea ice, thawing permafrost. Less extensive sea ice creates more open water, allowing stronger wave generation by winds, thus increasing wave-induced erosion along arctic shores. Therefore, the acceleration of erosion and thermo-abrasion of the coast can be caused by both increase of the air and water temperature and possible increasing of wind-wave activity.

Coastal zone degradation is an important problem for local population, industrial and transport system. On coast, built with disperse rocks of high ice content there are towns, pipelines and communications, navigation stations, etc. All these objects are in a danger of destruction. In last decade fast shore cliff retreating and activation of surface cryogenic events had frequently lead to collapses of houses, geodesy signs, navigation objects and so on.

Sea level uplifting causes negative cryogenic processes activation even far inshore: catastrophic development of ravines with thermal erosion, caverns of thermal karst and thermal suffusion, cryosolifluctive slope erosion. Because of spreading on vast area and high rate of extension, these processes may become hazard of more importance than ice-saturated cliffs retreating itself and may cause considerable damage, render null enterprise and extraction efficiency.

Man-caused factor accelerates dangerous morpholitodynamic processes: aeolian, thermal erosion and karst. This leads to decrease in natural shore stability and accelerates wash-away. Land reclaiming and cultivation had turned out to be destruction of existing morpholitodynamic system, and also had led to tangible economic losses.

Beginning in the nineteen-eighties, the Research Laboratory of Geoecology of the North (Faculty of Geography, MSU) has carried out permanent monitoring of the coastal dynamics and associated exogenous processes nearby Varandei Isl. (Ogorodov, 2004).

Active exploitation of the Varandei industrial area started in the seventies. Varandei Island was subjected to the strongest human impact. Here, the main industrial base was formed, and Novyi Varandei, a settlement for 3.5 thousand inhabitants, was built. The well-drained dune belt of the Holocene terrace (first morphogenetic complex), composed of sand beds with low ice content, was chosen as the place for the settlement, oil terminal and storehouses, because it seemed to be more stable from the engineering-geological point of view than the surrounding swampy tundra lowland (second morphogenetic complex).

Construction of the settlement and industrial base practically at the edge of the abrasion cliff demanded repeated withdrawals of sand and sand-pebble sediments from the avandune and beach.

Within the zone of industrial exploitation, the coastal bluff and the coastal zone experienced considerable mechanical deformations of the landforms because of transport ramps, mechanical leveling of coastal declivities and other human disturbances. Uncontrolled use of transport and construction techniques including caterpillars caused degradation of soil and plant covers of the whole dune belt of Varandei Island. Under conditions of deep seasonal melting, the dune belt formed of fine sands is subjected to deflation and thermoerosion. The extent and rate of these processes has been so great that in places the surface of the island became 1-3 m lower than before the period of exploitation. Deflation hollows became widespread. Numerous deflation-thermoerosional gullies were formed in the abrasion cliff. As a result, the cliff became lower, its homogeneity was disturbed, the amount of sediments supplied to the coastal zone decreased and, finally, the coasts became less stable, and the rate of retreat increased.

Coastal protection in the area close to the Novyi Varandei settlement (the region of wave energy flow divergence and, correspondingly, sediment flow formation) caused a decrease in sediment supply to the adjacent areas and, hence, their erosion.

After an earth-dam and a bridge were constructed in the eastern part of the Varandei Island, the height of storm surges increased. The latter is an important factor of coastal dynamics. Previously, during high surges corresponding in time with tides, water was partly flowing into the branches and channels, thus lowering the surge height and decreasing its influence upon the coast.

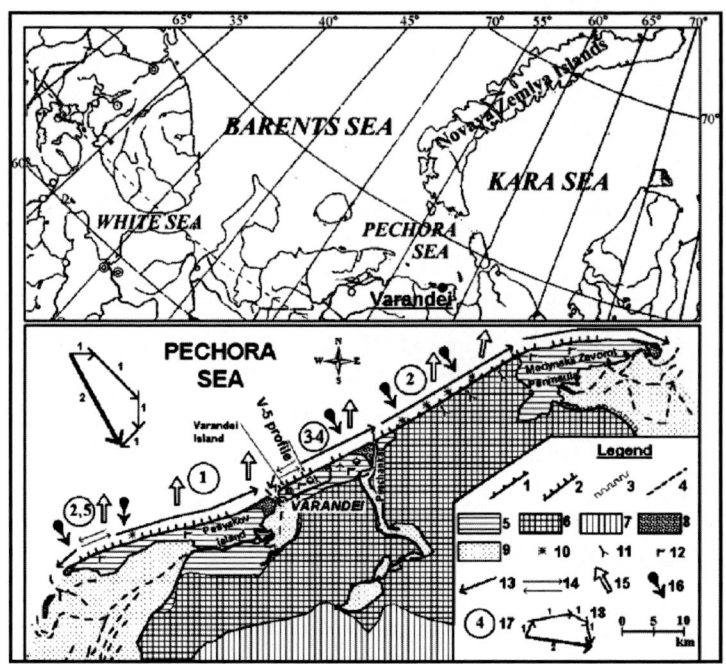

Figure. 3. Geomorphology of Varandei area.

Under the existing conditions of intensive human impact, the coastal erosion rate increased considerably in the mid to late seventies. In some years in some sites it was up to 7-10 m/year. The rate of coastal retreat slightly decreased, down to 1.5-2 m/year, after the coastal-protection construction was built near the Novyi Varandei settlement. However, it remained high in the adjacent areas. Recent measurements during 1987-2000 have shown that the rate of coastal retreat in the region around the settlement increased and reached 3-4 m/year: that is twice as high as in the regions that are not affected by human activity (Figure. 3).

Key: *Types of abrasion coasts*: 1 – with thermoabrasion or abrasion-thermodenudation cliff in dense boulder loams; 2 – with wave-cut cliff in sand and peat beds with low ice content; 3 – dead cliffs. *Elements of bottom relief*: 4 – big channels of subaerial and hydrogenic origin. *Types of terrestrial relief*: 5 – Holocene marine transgressive terrace with dune belt (up to 5-12 m asl) in the frontal part and laida (up to 2.5-3 m) in the inner part; 6 – Late Pleistocene-Holocene alluvial-lacustrine terrace up to 5-15 m high with thermokarst dissection; 7 – Middle Pleistocene glacial(ice?)-marine denudation plain (above 20 m high) with erosional dissection; 8 – Holocene free accumulative forms (beaches with well-developed profiles). *Elements of morpholithodynamics*: 9 – areas of lagoon accumulation within tidal flats and bays; 10 – "**clayey bench**"; 11 – regions of active gully thermoerosion; 12 – regions of active deflation; 13 – average multi-annual directions of sediment flows; 14 – areas of bilateral sediment flows; 15 – removal of fine-grained material along small discharge channels; 16 – release of the rock debris and pebbles from submarine coastal slope; 17 – measured average multi-annual rate of coastal bluff retreat, m/year; 18 – energetic polygon plotted on the basis of hydro-meteostation Varandei data, where (1) – rhumb component of the wave energy flow; (2) – wave energetic resultant, 1 mm of the arrow length = 1 arbitrary unit of wave energy.

At present time the situation on Varandei island is near critical level. The reclamation turned out to be not only the destruction of natural coastal system, but have caused economic losses. These losses will grow from year to year as the coast cliff will draw towards settlement centre. The petroleum base is under the immediate threat: in July 2002 the distance from cliff break to the nearest reservoir was less than 6m.

Recent activation of industrial activity in Pechora region requires important decisions on places for new buildings and reclamation strategy. The negative example of Varandei district urges the necessity of competent and ecologically-based approach to reclamation of new coastal areas.

Man created processes due to natural disbalance of coastal systems. This activation leads to essential increase in rates of shore and bottom abrasion. When pipes are paved through areas with active erosion, they may be damaged by possible exposure, slackening and mechanical deformations. Embedding, landfilling and other protection measures may be of low effect because areas with active erosion are characterized by routed abrasion of drain areas and underwater coast slope. After exposure, pipes are open to direct dynamic affection of sea ice. Stamping of shore ice causes bulk formation, which may lead to destruction of coastal buildings and pipes.

For reduction of ecological hazards the constant monitoring of shore dynamics is recommended, as well as prospecting and organization of environmentally safe extraction of construction materials, and compliance with engineering and ecological regulations established. Objective estimation, analysis and forecasting of technogenic factor affection on coastal systems may considerably decrease hazards and fulfill safe use of engineering works.

CONCLUSION

Coastal development is controlled by interrelated global climatic, geological-geomorphological and hydrophysical processes. Due to this fact, new and complex approaches for complicated and interdependent coastal processes are required. Coastal zone of every country is mostly a zone of enhanced economic interest. Algorithm for coastal zone development as a function of climatic zone is the base for calculating of development scenarios. In different climatic zones possible climate warming and rising of ocean level will cause absolutely different consequences. This fact is due to dissimilarity in paleogeographic development, geological and geomorphological structure, contemporary evolution, biogeographic peculiarities, etc. Every climatic zone has its own dominating features and processes, which determine the main direction for natural processes in present and in future. That's why developing algorithm for all world oceans could not be the same.

Time distribution analysis of spatial structure of atmospheric pressure and temperature fields in 1950-2007 has shown that processes of warmth redistribution in near-surface layer over oceans and continents are inversed. For example, in last quarter of XX century (compare to 1950-1974) the near-surface temperature had an increase over continents, while over oceans and several seas the cooling of suprazone was clearly observed.

Negative anomalies of ocean temperature are surely connected with processes of vertical convection. These processes involve at least the whole ocean suprazone (-100m) to the ocean-atmosphere thermoexchange. This fact concerned, ocean heat elimination (similar to warming over continents) is to be quite possible. If so, the global warming is supposed to be connected with natural redistribution of heat between oceans and continents because of internal dynamics of climatic system.

Possible climate warming will cause in Arctic more significant, even catastrophic, environmental changes, which may occur earlier than in low latitudes. The irregular climate warming has led to decrease of Arctic drift ice. This had cause activation of storms, increase in seasonally thawed layer and speed-up of shore retreating. Comparison with paleoclimatic models shows that global warming should lead to environmental conditions similar to Riss-Wurm Interglacial, when Arctic ocean was practically free from pack ice.

The main feature of Arctic Ocean coastal zone is permafrost rocks. In case of sea level rising, Arctic coastal zone will suffer intensive erosion, which will affect bioproduction, greenhouse gases concentration (increase), river outflow, hydrogeological regime, living conditions, etc.

Every Arctic sea has its own peculiarities, both in past and future development of environment. We believe Norwegian and Barents seas to have minimal changes in shore and shelf zones in the XXI-st Century. Fundamental changes are supposed to occur in hydrological pattern. The major changes will take place in Laptev and East-Siberian seas. Storm frequency and ice-free time will increase, the last by 50% (from 2 to 3 months, on our opinion). Consequently, wave action on coastline and bottom in the XXI st Century will be intensified nearly by 1.5-2 times. This will deepen thermal abrasion of coasts so sediment supply (organic material as well) also will increase.

In East-Siberian Sea climate warming will cause the most significant environment changes, mainly in ice-cover (decrease), thermal abrasion (acceleration) and change of sediment regime in coastal zone and on the shelf.

Coastal zone degradation is an important problem for local population, industrial and transport system. On coast, built with disperse rocks of high stone ice percentage there are towns, pipelines and communications, navigation stations, etc. All these objects are in a danger of destruction.

Technogenic factor accelerates dangerous morpholitodynamic processes: aeolian, slope, thermal erosion and karst. This leads to decrease in natural shore stability and accelerates wash-away. Geoecological situation in areas of intensive reclamation in the Pecohora Sea coastal zone, especially Vsarandei settlement, is near critical. Land reclaiming and cultivation had turned out to be destruction of existing morpholitodynamic system, and also had led to tangible economic losses.

REFERENCE

Alekseev, G. V., Bulatov, L. V. & Zaharov, V. F. (2004). Changes in Atlantic water layer in XX century. *In: Formation and dynamics of contemporary Arctic climate.* Gycmeteoizdat; 232-249 (in Russian).

Are, F. E. (1980). *Termoabrasion of marine coasts.* Moscow, Nauka, 320 P. (in Russian).

Belov, N. A. (1966). Contemporary geological processes and sea bottom dynamics of Laptev sea. *Works of AARI, vol. 269*, 53-56 (in Russian).

Byshev, V. I. (2003). *Synoptic and large-scale variations of ocean and atmosphere.* Moscow: Nauka; 2003 (in Russian).

Byshev, V. I., Kononova, N. K., Neyman, V.G. & Romanov, U. A. (2002). Peculiarities of Northern Hemisphere climate dynamics in XX century. *RAS Reports, vol. 384, 5*, 674-681 (in Russian).

Byshev, V. I., Kononova, N. K., Neyman, V. G. & Romanov, U. A. (2004).

Quantitative estimate of climate variation parameters of oceea—atmosphere system. *Oceanology, vol. 44*, 2, 341-353 (in Russian).

Byshev, V. I., Oprove, N. K. & Navrockaya, S. E (1993). Anomal condition of Newfoundland energoactive zone in 1990. *RAS Reports, vol. 331*, 6, 735-738 (in Russian).

Byshev, V. I., Neyman, V. G. & Romanov, U. A. (2005). The inversion of climte change over Continents and oceans. *RAS Reports, vol. 400, 1*, 98-104 (in Russian).

Byshev, V. I., Neyman, V. G. & Romanov U. A. (2006). The substantial difference in large-scale near-surface temperature changes over oceans and continents. *Oceanology, vol. 46*, 2, 165-177 (in Russian).

Byshev, V. I., Neyman, V. G. & Romanov U. A. (2009). Natural factors of global variability of contemporary climate. *RAS New, Geography section, 1*, 7-13 (in Russian).

Gill, A. (1986). *Atmosphere and ocean dynamics.* Moscow, Mir (in Russian).

Grigoriev, M. N. (2008). Criomorphogenesis and lithodynamics of east Siberia coastal zone. *Abstract Dc of Sci work*, Yakutsk, 41 P (in Russian).

Zveryaev, I. I. & Arhipkin, A. V. (2008). Structure of surface temperature climatic variability og Mediterranean sea. Part 1, Standard fluctuations and linear trends. *Meteorology and hydrology, 6*, 55-64 (in Russian).

Ivanov, V. V. & Korablev, A. A. (2004). Atlantic waters in Arctic basin and Arctic seas. In: Formation and dynamics of contemporary Arctic climate. St.-Petersburg; *Hydrometeoizdat*; 208-231 (in Russian).

Klige, R. K., Danilov, I. D., Konischev & V.N. (1998). *Hydrosphere history.* Moscow, Nauchniy Mir (in Russian).

Monin, A. S. (1969). Weather forecasting as a physic problem. Moscow, Nauka (in Russian).

"MEGAPOLYGON Experiment" (1992). Moscow, Nauka (in Russian).

Ogorodov, S. A. (2004). Coastal dynamics in Varandei area (the Pechora Sea) under man-caused conditions, *Geoecology, № 3*, 273-278 (in Russian).

Chiang, J. C. H. & Kushnir, Yo. (2000). Interdecadal in eastern Pacific ITCZ variability and its influence on the Atlantic ITCZ. *Geoph. Res. Lett., V. 27, № 22*, 3687-3690.

Hasselmann, K. (1976). Stochastic climate models. *Part I. Theory. Tellus, V. 28, № 6*, 473-485.

Hasselmann, K. & Frankignoul, C. (1977). Stochastic climate models. Part II. Application to sea surface temperature anomalies and termocline variability. *Tellus, V. 29, № 4*, 289-305.

Jones, P. D. & Moberg, A. (2003). Hemisphere and large-scale surface air temperature variations: An extensive revision and an update to 2001. *J. Climate, V. 16*, 206-223.

Latif, M. (2001). Tropical Pacific/Atlantic ocean Interactions at Multy-Decadal Time scales. *Geoph. Res. Lett., V. 28, № 3*, 539-542.

Mo, K. C. & Hakkinen, S. (2001). Decadal Variations in the tropical South Atlantic and Linkages to the Pacific. *Geoph. Res. Lett., V. 28, № 10*, 2065-2068.

Chapter 5

ALPINE MEADOW ON THE TIBETAN PLATEAU WAS A CO_2 SINK IN PEAK GROWING SEASON REVEALED BY KEELING PLOT METHOD

Xiaoyong Cui[1,2], Hongchao Tan[1], Yibo Wu[1], Jing Wu[1], Yangong Du[1], Yongcui Deng[1] and Yanhong Tang[3]

[1]College of Resource and Environmental Science, Graduate University of the Chinese Academy of Sciences, Beijing 100049, P.R. China.
[2]Northwest Institute of Plateau Biology, the Chinese Academy of Sciences, Xining 810001, P.R. China.
[3]National Institute for Environmental Studies, Tsukuba 305-8506, Japan.

ABSTRACT

Alpine meadow occupies about 37% of the total 2.5 ×10^6 km^2 of the Tibetan Plateau. Gas exchange between this ecosystem and atmosphere may have a large impact on greenhouse gases budget of the whole plateau. It is still under debate whether alpine meadow is a sink or source of atmospheric CO_2 in growing season.

Air samples were taken along vertical profiles in the boundary layer above three communities: *Kobresia humilis* alpine meadow, *K. tibetica* swamp meadow, and *Potentilla fruticosa* shrub meadow, in growing season in 2007. Relative contribution of gross photosynthesis (GP) of the vegetation and gross respiration (R, including soil and plant respiration) to net ecosystem exchange of CO_2 was calculated by constructing Keeling plots.

GP was greater than R for *K. humilis* alpine meadow and *P. fruticosa* shrub meadow in clear days, indicating carbon sinks of these communities. The ratio of GP/R was higher in the former community than in the latter one, suggesting stronger sink for the *K. humilis* alpine meadow. GP/R decreased to near 1 when it was cloudy in these two ecosystems. During showers, the points deviated largely from the lines in Keeling plots. However, progressive transition between sink and source was discernable in these plots.

GP was roughly similar or slightly lower than R for the *K. tibetica* swamp meadow even in clear days. The swamp meadow accumulated deep peat layer during its development and may continue sequestrate C in the soil. Large quantity of exotic dissolved and particulate organic carbon transported from surrounding communities and from herd excreta in the swamp meadow were supposed to result in high gross respiration in the this ecosystem.

The results implied that the alpine meadow ecosystem on the Tibetan Plateau was the sink of atmospheric CO_2 in growing season.

Keywords: Meadow; shrub; wetland; carbon budget; gross photosynthesis; respiration.

INTRODUCTION

Atmospheric CO_2 plays an important role in climate change. Terrestrial ecosystems are essential in global carbon cycling and atmospheric CO_2 budget due to their huge C storage and C fluxes, as well as high sensitivity to climate change (IPCC, 2007). Recently, micrometeorological eddy covariance approaches have been rapidly developed and installed as nearly routine equipments in more and more field investigation sites (Wofsy et al., 1993; Baldocchi, 2003; Williams et al., 2004; Fu et al., 2006; Hirata et al., 2007; Pypker et al., 2007; Saito et al., 2009). These real time monitoring systems measure net ecosystem exchange (NEE) of CO_2, H_2O or other volatile chemicals, thus are valuable in clarifying carbon sink or source of ecosystems. However, they provide little information on relative contribution of the components, such as gross photosynthesis (GP) and respiration (R). Extrapolation of nighttime NEE to daytime R is a common way to estimate ecosystem R and consequently to reckon GP and GP/R, in spite of probably large deviation (Wohlfahrt et al., 2005). Therefore, partitioning NEE into GP and R by other means is of great importance to verify or complement NEE data and to clarify mechanisms controlling GP and R.

Isotope signature offers an effective measure to partition composite flux into components of known isotope composition or discrimination effect (Flanagan and Ehleringer, 1998; Fry, 2006; Bowling et al., 2008). Keeling plot affords a way to figure out isotope signature of the composite flux that is unknown or not able to directly measure (Keeling, 1958). Evapotranspiration and nocturnal respiration have been successfully divided into individual components, whereas separating GP with R is less effective using air CO_2 samples and ^{18}O performs better than ^{13}C (Yakir and Wang, 1996; Flanagan and Ehleringer, 1998; Yakir and Sternberg, 2000; Pataki et al., 2003; Williams et al., 2004; Pypker et al., 2007; Bowling et al., 2008).

Grasslands cover some 40 % of the earth's surface, excluding Greenland and Antarctic (Reyers et al., 2005). CO_2 flux measurements have been conducted in more and more grassland ecosystems due to their potential importance in global carbon budget and climate change (Zheng et al., 2000; Gilmanov et al., 2003; Xu and Baldocchi, 2004). Grasslands are also dominant vegetation types on the vast Tibetan Plateau, occupying 46 % of the over 2.5×10^6 km^2 area (Zheng et al., 2000; Gilmanov et al., 2003; Xu and Baldocchi, 2004). The high elevation ecosystems store a large amount of C in the belowground (Ni, 2002), which may release under climate change or human disturbance (Cao et al., 2004). Although recent eddy covariance measurements indicate that alpine grasslands on the Tibetan Plateau are currently acting as carbon sinks (Gu et al., 2003; Kato et al., 2004a; Zhao et al., 2005), opposite results by other methods also exist (Zhang et al., 2003; Yi and Yang, 2006).

Therefore, the aims of this study are to (1) partition GP and R of three dominant grasslands (an alpine meadow, an alpine shrub meadow and an alpine swamp meadow) using Keeling plot approach, and (2) determine if it is feasible to perform Keeling plot under unstable weather conditions when the eddy covariance measurements may not work properly.

MATERIALS AND METHODS

Study Site Description

Measurements were conducted in Haibei Alpine Meadow Ecosystem Research Station (lat. 37°29`~37°45` N, long. 101°12` ~101°23` E, 3250 m altitude) at the northeast edge of the Qinghai-Tibetan Plateau. The climate is dominated by the Southeast Monsoon and Siberia High Pressure System. It has a continental monsoon type climate, with severe and long winter and short cool summer. The mean annual air temperature was −1.7 ºC with extreme minimum of −37.1 ºC and maximum of 27.6 ºC. Annual precipitation, ranging from 426 to 860 mm, averaged 560 mm in the past 20 years, of which 85% was concentrated in growing season from May to early October. The annual average sunshine duration was 2462.7 h, about 60.1% of total available (Yi and Yang, 2006).

Three alpine vegetation types, a meadow, a shrub meadow, and a swamp meadow were studied. The meadow was located on a flat plain. It contained only a herbaceous layer, and was dominated by *Kobresia humilis, Polygonum viviparum, Carex atrofusca, Saussurea superba, Elymus nutans*, and *Gentiana straminea*. Other common species included *Potentilla saundersiana, Leontopodium nanum, Lancea tibetica, Festuca ovina, Festuca rubra, Stipa aliena, Elymus nutans, Helictotrichon tibetica, Koeleria cristata, and Poa crymophila*. Vegetation coverage ranged from 75% to 80%, of which 98% by grasses and 2% by bare soil. The maximum aboveground biomass varied within the range of 342 ± 50 g DW m^{-2} (average ± S.D.). The LAI (leaf area index) reached a maximum of about four in July.

The shrub meadow had two vegetation layers, a shrub layer dominated by *Potentilla fruticosa* and a herbaceous layer dominated by *F. rubra, Stipa alpine, K. humilis, E. nutans, Polygonum viviparum, Ranunculus indivisus, Poa pratensis* and *P. saundersiana. Salix oritrepha, Sibiraea artgustata, Spiraea alpine, Garagana jubata* interspersed in shrub layer. Vegetation coverage ranged from 60% to 75%, of which 50% by shrubs, and 48% by grasses and 2% by bare soil (Zhou and Wu, 2006). Generally there was a litter layer of about 3 cm thickness under the shrub clusters. Alpine grasses were about 25-30 cm high, while alpine shrubs were 50-70 cm high.

The swamp meadow was dominated by *Kobresia tibetica. Blysmus sinocompressus* and *Carex pamirensis* Marsh were the sub-dominant species, and *Carex atrofusca, Ligularia virgaurea, Hippuris vulgaris, Pediculars longiflora, Swertia przewalskii* were also common. Around the edge of the wetland, *Saussurea stilla, Carex moorcropt, Kobresia tibetica*, and *K. humilis* became abundant around the edge. Vegetation height varied from 10 to 50 cm.

The soils developed in the *K. humilis* meadow, the *P. fruticosa* shrub meadow, and *K. tibetica* swamp meadow were Mat-Cryic Cambisol, Mol-Gryic Cambisol, and Organic Cryic Gleysols (Chinese Soil Taxonomy Research Group, 1995), corresponding to Gelic Cambisol (WRB, 1998). They were rich in organic carbon and had an Udic soil moisture regime.

The *K. humilis* meadow had been used as winter pasture from late September to the end of next April since 1982. The shrub meadow and swamp meadow were grazed by yaks and sheep in summer and autumn.

Sample Collection

Air samples were manually collected from July 30 to August 1, and on August 15, 2007. Vertical poles with height marks were installed in each site. Samples were gathered at four heights, i.e. 2, 3, 4, and 5 m above canopy at each site. Aluminium foil compound bags (Teldar®, Delin Gas Packing Co., Ltd, Dalin, China) of 1 L in volume were used to gather and temporarily store air samples. Teflon tubes were connected to the inlet valves of the bags during sample collection. The valves switched to close upon completion of sampling. Syringe samplers were then used to take air out of the bags for measurement.

To minimize disturbance by people, one bag was put in a sealed box before sampling. Teflon tube to the bag was extended for 15 cm out of the box. A long plastic tube was inserted into the box. Air inside of the box was drawn out through the hard plastic tube by a pump. The negative pressure in the box then sucked air into the bag in the box. It took roughly 2 min to collect one sample, and less than 10 min to complete sampling at one site.

At alpine meadow site, samplings were conducted for 6 times to cover a range of daytime as well as weather condition. At shrub meadow and swamp meadow, air samples were taken one time only at clear day.

Plant and soil samples were taken after air sampling during the clear days. At each site, three plots of 25 cm ×25 cm were randomly selected. Plant aboveground living parts were harvested and separated into leaves and other components. In shrub meadow site, leaves of *P. fruticosa* were taken from several clusters. Soil samples were taken only to 10 cm depth. These samples were oven-dried and ball-milled finely for isotope analysis.

Measurements

CO_2 concentration was measured by HP6890 gas chromatograph (Agilent Co.) system with flame ionization detector (FID). Temperature in Ni-catalyst oven and FID was 375 °C and 220 °C, respectively. Injection/detection and column (stainless steel SS-3 m×2 mm× Porapak Q) oven temperature was 55 °C and 330 °C, respectively. Ultra pure N_2 was used as carrier gas with a flow rate of 30 ml min^{-1}. A certified CO_2 standard gas of 456×10^{-9} L L^{-1} (China National Research Center for Certified Reference Materials, Beijing) was used for calibration.

Carbon isotope composition of air, plant, and soil samples was determined by Thermo Finnigen MAT DELTA plus XP (Thermo Finnigen, Bremen, Germany) isotope ratio mass spectrometer operating in continuous flow mode. Carbon isotope ratio was presented in δ (‰) notation:

$$\delta = [(R_{sample} - R_{standard}) / R_{standard}] \times 1000 \tag{1}$$

where R_{sample} and $R_{standard}$ were $^{13}C/^{12}C$ ratio of the sample and the V-PDB standard. Precision of duplicate measurements was 0.1% for $\delta^{13}C$.

Calculation

Net CO_2 flux from vegetation-soil system (v) was partitioned into gross photosynthesis (GP) and respiration (R) by Keeling plot method (Yakir and Wang, 1996). This approach supposed that at any height z in the boundary layer above canopy, the air was a mixture of background atmosphere air (b) and CO_2 added or removed by vegetation-soil system. Concentration of CO_2 was calculated according to:

$$C_z = C_b + C_v \tag{2}$$

CO_2 mixing followed the rule of isotope mass balance, which could be approximately expressed by the equation below:

$$C_z \delta_z = C_b \delta_b + C_v \delta_v \tag{3}$$

where δ_v was the carbon isotope composition of the virtual mixture of CO_2 respired from vegetation-soil system with CO_2 fixed by photosynthesis. Equations (2) and (3) were rearranged into:

$$\delta_z C_z = C_b (\delta_b - \delta_v) + \delta_v C_z \tag{4}$$

By plotting $\delta_z C_z$ versus C_z, δ_v could be obtained as the slope of the fitted line (Miller and Tans, 2003). The contribution of component GP and R to net CO_2 flux was then calculated by:

$$f_{GP} = (\delta_{GP} - \delta_v)/(\delta_{GP} - \delta_R) \tag{5}$$

and

$$f_R = 1 - f_{GP} \tag{6}$$

where δ_{GP} and δ_R were isotope signature of CO_2 associated with photosynthesis and respiration from vegetation-soil system. The values were estimated from leaf $\delta^{13}C$ and from average $\delta^{13}C$ of plant and soil organic matter (Yakir and Wang, 1996).

RESULTS

Under clear condition, the concentration and carbon isotope composition of CO_2 in the atmosphere above canopy showed smooth and inverse vertical gradients (Figure 1). By Keeling plot in Miller/Tans style (Figure 2), δ_v was obtained for the three vegetation types. Gross photosynthesis and respiration was then partitioned based on the measured plant and soil carbon isotope signatures (Table 1). The *K. humilis* meadow showed higher ratio of GP/R than *P. fruticosa* shrub meadow and *K. tibetica* swamp meadow under clear sky. The swamp meadow was net carbon source and the other two vegetation types were carbon sink.

When the weather was not stable, vertical profiles became unsmooth for both CO_2 concentration and its carbon isotope signature above canopy (Figure 3). However, linear relationship maintained between reciprocal of CO_2 concentration and its $\delta^{13}C$ value (Figure 4).

DISCUSSION

Carbon Sink or Source for the Three Alpine Meadow Ecosystems

Using Keeling plot method, the current research demonstrated that gross photosynthesis was larger than ecosystem respiration in *K. humilis* meadow under clear condition in peak growing period. The ratio of GP/R decreased a little from early in the morning to late afternoon (Table 1), however, it was still net carbon absorption during the whole day. This result was consistent with other studies in similar vegetation types on the Tibetan Plateau (Gu et al., 2003; Kato et al., 2004a; Kato et al., 2004b; Zhao et al., 2005). The GP/R was calculated to be around 2.0 in peak growing season by eddy covariance method (Gu et al., 2003), which was very close to our estimation with Keeling plot approach (Table 1).

Long term eddy covariance observation data indicated that the *K. humilis* meadow was a stronger carbon sink than the *P. fruticosa* shrub meadow, and the *K. tibetica* swamp meadow was net carbon source (Zhao et al., 2005). The current short term study was in accordance with the above result (Table 1). The *P. fruticosa* shrub meadow generally occupied southward slopes. Therefore, it had shallower soil depth and poorer soil moisture than the *K. humilis* meadow. Besides, plant photosynthetic capacity was lower in shrubs than in grasses and herbs on the Tibetan Plateau (He et al., 2006). Photosynthetic depression under high light and high temperature may also contribute to the lower GP/R in this study (Fu et al., 2006). Depression of photosynthesis under high light was observed. Therefore, the shrub meadow had lower carbon sequestration capacity than the *K. humilis* meadow. The *K. tibetica* swamp meadow had thick peat layer and much lower aboveground biomass, with a maximum of about 330 g DW m^{-2} (Li et al., 2007; Zhang et al., 2008) in contrast to about 6222 g DW m^{-2} of a reed wetland in Denmark (Brix et al., 2001). Intensive yak grazing reduced photosynthetic leaf area and input large amount of dissolved organic carbon (DOC). DOC also fluxed in with water, and may play a role in large release of CO_2 in the swamp meadow.

Using biomass monitoring data and measured soil respiration by close chamber method, Zhang et al. (2003) suggested that the *K. humilis* meadow was a carbon source, with net release of 199 g C m^{-2} a^{-1}, probably due to an underestimation of NPP (Kato et al., 2004a). With a simple mass balance model considering CO_2 concentration and its carbon isotope signature in the air 2 m above ground in one day, Yi and Yang (2006) calculated that the *K. humilis* meadow was carbon source of about 33 g C m^{-2} a^{-1}, some 3 % of NPP. Since both photosynthesis and respiration were highly variable, GP/R may change greatly seasonally or even in a day (Gu et al., 2003; Kato et al., 2004b; Fu et al., 2006; Hirata et al., 2007; Saito et al., 2009). Consequently, an ecosystem may convert from carbon sink to source or vice versa instantaneously or seasonally. Therefore, measurement of $\delta^{13}CO_2$ for only one day was not sufficient to figure out annual carbon budget, despite of the lower $\delta^{13}CO_2$ value that implied higher R than GP at that time (Yi and Yang, 2006). In addition, the equations in their article were not correctly deduced.

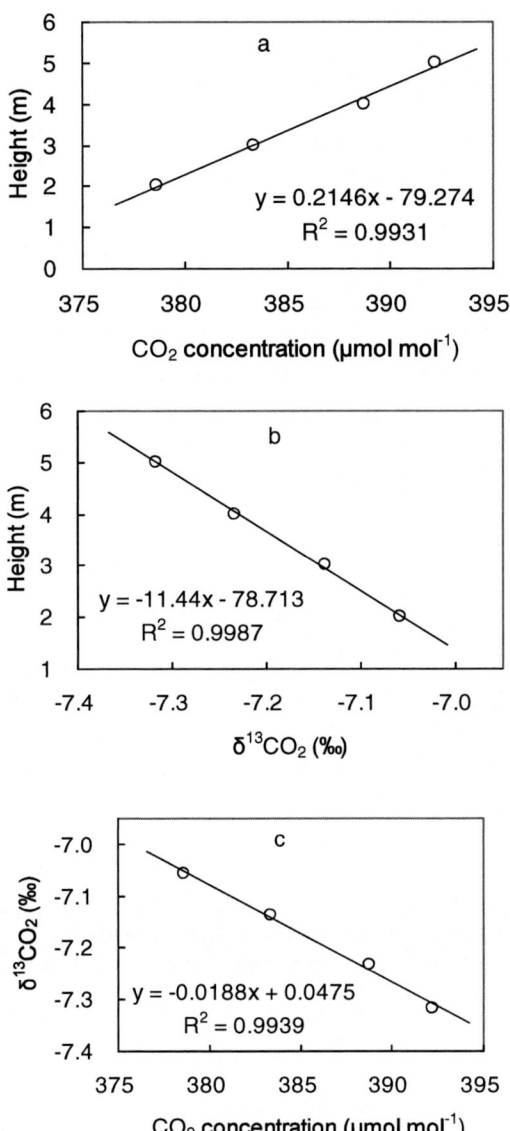

Figure 1. Vertical gradient of CO_2 concentration (a) and its carbon isotope composition (b) in the atmosphere above alpine *K. humilis* meadow canopy under clear condition at 16:40 (local time) on July 30, 2007. Relationship between CO_2 concentration and its carbon isotope composition was also shown in diagram (c). The patterns were similar for *K. humilis* meadow at other time and for *P. fruticosa* shrub meadow and *K. tibetica* swamp meadow under clear condition.

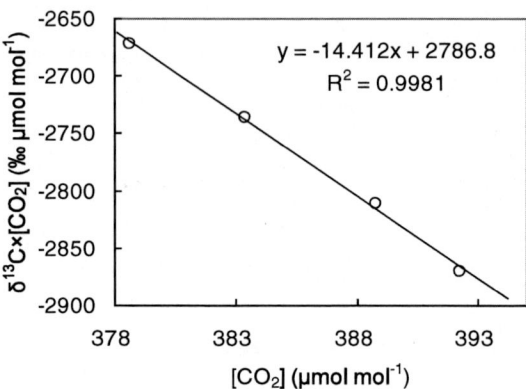

Figure 2. Keeling plot analysis to obtain carbon isotope signature of CO_2 flux from vegetation-soil system. The plot was in Miller/Tans style.

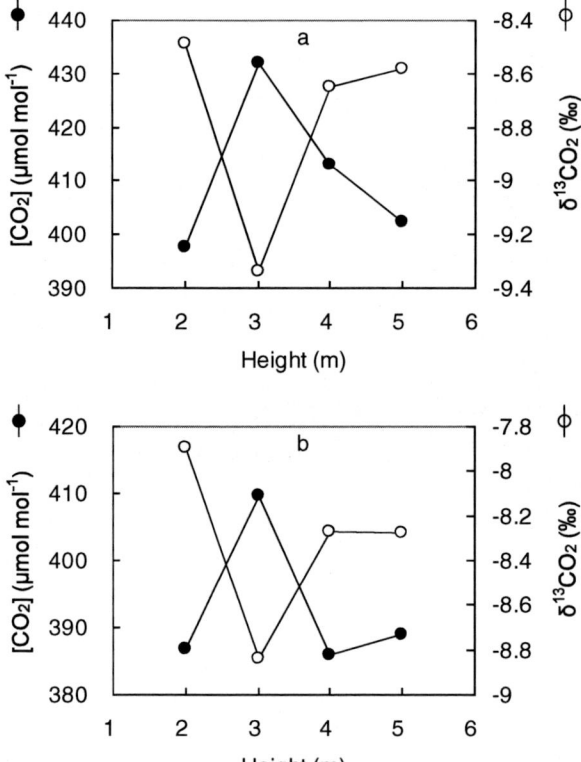

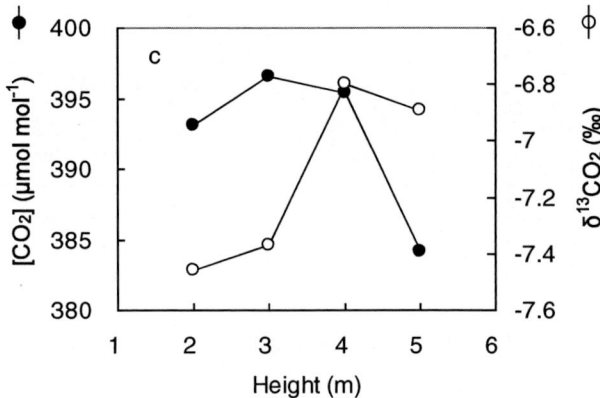

Figure 3. Vertical profiles of CO_2 concentration and its carbon isotope signature in the air above the canopy of *K. humilis* meadow at unsteady weather conditions. The samples were taken in different time as shown in Table 1.

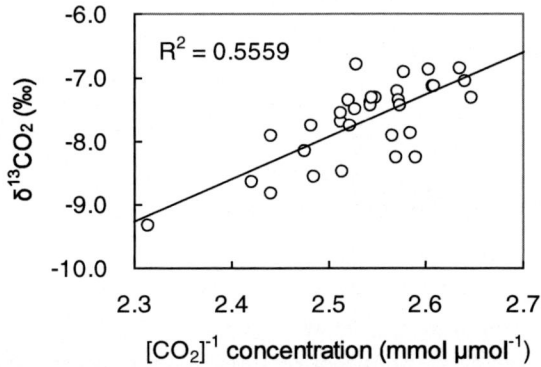

Figure 4. Relationship between CO_2 concentration and its carbon isotope signature analyzed with all pooled data.

Table 1. Ratio of gross photosynthesis and respiration in three alpine vegetation types under clear and shower weather conditions. Negative signal meant that gross photosynthesis and respiration was in opposite direction in net ecosystem exchange.

Vegetation type	GP/R	Weather	Time of the day
K. humilis meadow	-2.9	Sunshine	09:20
	-2.7	Sunshine	13:50
	-2.1	Sunshine	16:40

Table 1. (Continued)

Vegetation type	GP/R	Weather	Time of the day
P. fruticosa shrub meadow	-1.2	Sunshine	15:20
K. tibetica swamp meadow	-0.7	Sunshine	11:30
K. humilis meadow	3.4	Shower	17:00
	0.3	Shower	11:00
	1.2	Shower	14:20

Comparison of Three Approaches Involving Carbon Isotope to Estimate GP/R

Yi and Yang (2006) proposed a simple method to determine ecosystem CO_2 flux based on isotope mass balance. It assumed that imbalance of CO_2 uptake and release caused change of carbon isotope signature. Supposing ecosystem CO_2 flux was constrained in the boundary layer, we got the following equations:

$$(R_a - R_b) \times (x - y) + y \times R_a \times (\alpha_{plant}^{-1} - 1) = x \times R_s \quad (7)$$

where R denoted $^{13}C/^{12}C$; subscript a, b, and s meant air above canopy, air background, sources; x and y were annual flux of sources and sinks; $\alpha_{plant} = (\Delta+1) = (\delta_a +1)/(\delta_{plant} +1)$. The ratio GP/R was then obtained as:

$$GP/R = y/x = (\delta_{plant} - \delta_b)/(\delta_a - \delta_b - \delta_s - 1) \quad (8)$$

The formula was quite simple. It required the measurement of carbon isotope composition of only five components and without information on concentrations. However, the underlying assumption made it difficult to collect air samples. The idea way was to continuously sample the air throughout the year and calculate the averaged δ_a. The estimation may suffer from large bias with samples taken in short-term.

Keeling plot approach was powerful in partitioning respiration components or evapotranspiration (Yakir and Wang, 1996; Tu et al., 2001; Pataki et al., 2003; Carleton et al., 2004; Williams et al., 2004; Pypker et al., 2007). Sometimes, however, it failed to separate gross photosynthesis with respiration (Yakir and Wang, 1996). Several factors impaired the capability of partition GP from R by Keeling plot method. First, a difference in $\delta^{13}CO_2$ between GP and R was a prerequisite for partitioning. But it was generally small, especially in a community with multiple species (Yakir and Wang, 1996). Second, there was large temporal, spatial, and interspecies variation in carbon isotope signature for both GP and R. Isotope discrimination controlled $\delta^{13}C$ of photosynthetically fixed CO_2, and was sensitive to factors affecting stomatal conductance and photosynthetic activity. Soil and weather conditions may alter $\delta^{13}C$ of respired CO_2 (Bowling et al., 2002; Shim et al., 2009). Photosynthesis provided substrate to ecosystem respiration and also caused variation of respired $\delta^{13}CO_2$ (Bahn et al., 2009). Usually $\delta^{13}C$ of plant and soil organic matter was adopted as carbon isotope signature of respired CO_2. This sometime led to a deviation of up to 6 ‰ (Bowling et al., 2008). Third, a narrow range of CO_2 concentration among the samples resulted in large error in component partition (Pataki et al., 2003).

Since isotope signature for both GP and R was sensitive to biotic and abiotic factors, Keeling plot method may lose its power under unstable weather conditions, as shown in current study (Table 1, Figure 3). Nevertheless, the dynamic process of air CO_2 concentration and its isotope composition could be recorded if sampling in sufficiently high speed. We took one sample at a height for less than 2 min. The gradual change of CO_2 concentration and $\delta^{13}CO_2$ was roughly portrayed (Figure 3). Novel methods of online isotope measurement would be able probe into the fast transition processes and trace the ratio GP/R instantaneously (Tu et al., 2001; Griffis et al., 2004; Bickford et al., 2009).

Unlike additive components, photosynthesis removed CO_2 from respired CO_2. Therefore, partitioning GP and R by Keeling plot method was not straightforward. This resulted in two problems: (1) it became invalid when $\delta^{13}CO_2$ for GP and R was identical; (2) error caused by continuous change of $\delta^{13}CO_2$ even when GP equaled R. To address this issue, respiration and photosynthesis was considered as separately consequent processes. CO_2 concentration and its isotope composition in the air above canopy were calculated step by step according to mass balance principle, taking into account isotope mixing during respiration and discrimination associated with photosynthesis (Fry, 2006). First considering respiration at time t and after an interval t+1:

$$C_{t+1} = C_t + C_R \qquad (9)$$

$$\delta_{t+1}C_{t+1} = \delta_t C_t + \delta_R C_R \qquad (10)$$

where C was the concentration of exchanged CO_2. Assuming that photosynthesis occurred immediately at time t +1:

$$C_{t+2} = C_{t+1} - C_{GP} \tag{11}$$

$$\delta_{t+2}C_{t+1} = \delta_{t+1}C_{t+1} + \Delta(1 - C_{t+2}/C_{t+1}) \tag{12}$$

These equations could be rearranged into:

$$C_{fz} = C_i + C_{Rz} - C_{GPz} \tag{13}$$

$$\delta_{fz} = (\delta_i C_i + \delta_R C_{Rz} + \Delta C_{GPz})/(C_i + C_{Rz}) \tag{14}$$

where subscript z denoted height z aboveground, f and i was air at certain height and background air; $\Delta = (\alpha_{plant} - 1)$. Solving equations (13) and (14), we got:

$$C_{Rz} = \frac{(\delta_{fz} - \delta_i - \Delta)C_i + \Delta C_{fz}}{\delta_R - \delta_{fz} + \Delta} \tag{15}$$

$$C_{GPz} = C_i - C_{fz} + \frac{(\delta_{fz} - \delta_i - \Delta)C_i + \Delta C_{fz}}{\delta_R - \delta_{fz} + \Delta} \tag{16}$$

Taking the measured CO_2 concentration and its isotope signature at two heights, vertical difference in C_R and C_{GP} was calculated. GP/R was then computed as:

$$GP/R = (C_{GP1} - C_{GP2})/(C_{R1} - C_{R2}) \tag{17}$$

CO_2 concentration and its $\delta^{13}C$ were required for calculation. Besides, it also demanded photosynthetic carbon isotope discrimination factor and isotope signature of respired CO_2. Sensitivity analysis indicated that the calculated GP/R was sensitive to the above parameters. Therefore, the methods may be feasible for unique vegetations with careful determined parameters, even though it had a solid theoretical basis.

CONCLUSION

Under clear sky, the vertical profiles were smooth for the concentration and its C isotope signature of CO_2 in the air above canopies of three alpine meadows. Keeling plot method was capable of calculating δ value of the virtual mixture of CO_2 associated with gross photosynthesis and respiration. Relative contribution of GP and R to NEE could be partitioned. The *K. humilis* meadow and *P. frutisoca* shrub meadow were carbon sink and the *K. tibetica* swamp meadow was carbon source in peak growing season.

Keeling plot method produced unreasonable isotope signature using all the points in the vertical profile when it was cloudy or showery. Nevertheless, it could track the rapid change of boundary layer CO_2 concentration and its isotope composition. With fast online isotope measurement instrument, this method may be coupled with eddy covariance flux measurement system to partition GP and R even under versatile weather conditions.

ACKNOWLEDGMENT

This work was financially supported by the Natural Science Foundation of China (30670318), Ecological Recovery and Directional Management Technology in the program of forest science (2006BAD03A0402), and CAS grants (kzcx2-yw-418-02 and kzxx2-06-01).

REFERENCES

Bahn, M., Schmitt, M., Siegwolf, R., Richter, A. & Brüggemann, N. (2009). Does photosynthesis affect grassland soil-respired CO_2 and its carbon isotope composition on a diurnal timescale? *New Phytologist*, 182, 451-460.

Baldocchi, D. (2003). Assessing the eddy covariance technique for evaluating carbon dioxide exchange rates of ecosystems: past, present and future. *Global Change Biology*, 9, 479-492.

Bickford, C. P., Mcdowell, N. G., Erhardt, E. B. & Hanson, D.T. (2009) High-frequency field measurements of diurnal carbon isotope discrimination and internal conductance in a semi-arid species, Juniperus monosperma. *Plant, Cell and Environment*, *DOI*, 10.1111/j.1365-3040.2009.01959.x.

Bowling, D. R., McDowell, N. G., Bond, B. J., Law, B. E. & Ehleringer, J. R. (2002). ^{13}C content of ecosystem respiration is linked to precipitation and vapor pressure deficit. *Oecologia, 131*, 113-124.

Bowling, D. R., Pataki, D. E. & Randerson, J. T. (2008). Carbon isotopes in terrestrial ecosystem pools and CO_2 fluxes. *New Phytologist, 178*, 24-40.

Brix, H., Sorrel, B. K. & Lorenzen, B. (2001). Are Phragmites-dominated wetlands a net source or net sink of green gases? *Aquatic Botany, 69*, 313-324.

Cao, G., Tang, Y., Mo, W., Wang, Y., Li, Y. & Zhao, X. (2004). Grazing intensity alters soil respiration in an alpine meadow on the Tibetan plateau. *Soil Biology and Biochemistry, 36*, 237-243.

Carleton, S. A., Wolf, B. O. & de Rio, C. M. (2004). Keeling plots for hummingbirds: a method to estimate carbon isotope ratios of respired CO_2 in small vertebrates. *Oecologia, 141*, 1-6.

Chinese Soil Taxonomy Research Group. (1995). *Chinese Soil Taxonomy*. China Agricultural Science and Technology Press, Beijing, China.

Flanagan, L. B. & Ehleringer, J. R. (1998). Ecosystem-atmosphere CO_2 exchange: interpreting signals of change using stable isotope ratios. *Trends in Ecology and Evolution, 13*, 10-14.

Fry, B. (2006). Stable Isotope Ecology. Springer, New York, NY, USA

Fu, Y. L., Yu, G. R., Sun, X. M., Li, Y. N., Wen, X. F., Zhang, L. M., Li, Z. Q., Zhao, L. & Hao, Y. B. (2006). Depression of net ecosystem CO_2 exchange in semi-arid *Leymus chinensis* steppe and alpine shrub. *Agricultural and Forest Meteorology, 137*, 234-244.

Gilmanov, T., Verma, S., Sims, P., Meyers, T., Bradford, J., Burba, G. & Suyker, A. (2003). Gross primary production and light response parameters of four Southern Plains ecosystems estimated using long-term CO_2-flux tower measurements. *Global Biogeochemical Cycles, 17*, 1071, doi: 10.1029/2002GB002023.

Griffis, T. J., Baker, J. M., Sargent, S. D., Tanner, B. D. & Zhang, J. (2004). Measuring field-scale isotopic CO_2 fluxes with tunable diode laser absorption spectroscopy and micrometeorological techniques. *Agricultural and Forest Meteorology*, 124:15-29.

Gu, S., Tang, Y., Du, M., Kato, T., Li, Y., Cui, X. & Zhao, X. (2003). Short-term variation of CO_2 flux in relation to environmental controls in an alpine meadow on the Qinghai-Tibetan Plateau. *Journal of Geophysical Research, 108*, D21, 4670.

He, J.-S., Wang, Z., Wang, X., Schmid, B., Zuo, W., Zhou, M., Zheng, C., Wang, M. & Fang, J. (2006). A test of the generality of leaf trait relationships on the Tibetan Plateau. *New Phytologist, 170,* 835-848.

Hirata, R., Hirano, T., Saigusa, N., Fujinuma, Y., Inukai, K., Kitamori, Y., Takahashi, Y. & Yamamoto, S. (2007). Seasonal and interannual variations in carbon dioxide exchange of a temperate larch forest. *Agricultural and Forest Meteorology, 147,* 110-124.

IPCC, 2007. Climate Change (2007). *The Physical Science Basis.* Cambridge University Press, Cambridge, UK and New York, NY, USA.

Kato, T., Tang, Y., Gu, S., Cui, X., Hirota, M., Du, M., Li, Y., Zhao, X. & Oikawa, T. (2004a). Carbon dioxide exchange between the atmosphere and an alpine meadow ecosystem on the Qinghai-Tibetan Plateau, China. *Agricultural and Forest Meteorology, 124,* 121-134.

Kato, T., Tang, Y., Gu, S., Hirota, M., Cui, X., Du, M., Li, Y., Zhao, X. & Oikawa, T. (2004b). Seasonal patterns of gross primary production and ecosystem respiration in an alpine meadow ecosystem on the Qinghai-Tibetan Plateau. *Journal of Geophysical Research, 109,* D12109, doi:12110.11029/12003JD003951.

Keeling, C. (1958). The concentration and isotopic abundance of atmospheric carbon dioxide in rural areas. *Geochimica Cosmochimica Acta, 13,* 322-334.

Li, Y. N., Zhao, L., Zhao, X. Q., Wang, Q. X. & Zhang, F. W. (2007). The features of soil organic matters supplement and CO_2 exchange between ground and atmosphere in alpine wetland ecosystem. *Journal of Glaciology and Geocryology, 29,* 940-946.

Miller, J. B. & Tans, P. P. (2003). Calculating isotopic fractionation from atmospheric measurements at various scales. *Tellus, 55B,* 207-214.

Ni, J., (2002). Carbon storage in grasslands of China. *Journal of Arid Environments, 50,* 205-218.

Pataki, D. E., Ehleringer, J. R., Flanagan, L. B., Yakir, D., Bowling, D. R., Still, C. J., Buchmann, N., Kaplan, J. O. & Berry, J. A. (2003). The application and interpretation of Keeling plots in terrestrial carbon cycle research. *Global Biogeochemical Cycles, 17,* 1022, doi:1010.1029/2001 GB001850.

Pypker, T., Unsworth, M., Mix, A., Rugh, W., Ocheltree, T., Alstad, K. & Bond, B. (2007). Using nocturnal cold air drainage flow to monitor ecosystem processes in complex terrain. *Ecological Applications, 17,* 702-714.

Reyers, B., Nel, J., Egoh, B., Jonas, Z. & Rouget, M. (2005). National Grasslands Biodiversity Program: Grassland Biodiversity Profile and Spatial Biodiversity Priority Assessment No. (CSIR Report Number) ENV-S-C 2005-102, CSIR.

Saito, M., Kato, T. & Tang, Y. (2009). Temperature controls ecosystem CO_2 exchange of an alpine meadow on the northeastern Tibetan Plateau. *Global Change Biology, 15*, 221-228.

Shim, J. H., Pendall, E., Morgan, J. A. & Ojima, D.S. (2009). Wetting and drying cycles drive variations in the stable carbon isotope ratio of respired carbon dioxide in semi-arid grassland. *Oecologia, 160*, 321-333.

Tu, K. P., Brooks, P. D. & Dawson, T. E. (2001). Using septum-capped vials with continuous-flow isotope ratio mass spectrometric analysis of atmospheric CO_2 for Keeling plot applications. *Rapid Communications in Mass Spectrometry, 15*, 952-956.

Williams, D. G., Cable, W., Hultine, K., Hoedjes, J. C. B., Yepez, E. A., Simonneaux, V., Er-Raki, S., Boulet, G., Bruin, H. A. R., Chehbouni, A., Hartogensis, O. K. & Timouk, F. (2004). Evapotranspiration components determined by stable isotope, sap flow and eddy covariance techniques. *Agricultural and Forest Meteorology, 125*, 241-258.

Wofsy, S., Goulden, M., Munger, J. Fan, S.-M., Bakwin, P. S., Daube, B. C., Bassow, S. L. & Bazzaz, F. A. (1993). Net exchange of CO_2 in a mid-latitude forest. *Science, 260*, 1314-1317.

Wohlfahrt, G., Bahn, M., Haslwanter, A., Newesely, C. & Cernusca, A. (2005). Estimation of daytime ecosystem respiration to determine gross primary production of a mountain meadow. *Agricultural and Forest Meteorology, 130*, 13-25.

WRB. (1998). 1998 WRB, World Reference Base for Soil Resources. Food and Agriculture Organization of the United Nations, Rome.

Xu, L. & Baldocchi, D., (2004). Seasonal variation in carbon dioxide exchange over a Mediterranean annual grassland in California. *Agricultural and Forest Meteorology, 123*, 79-96.

Yakir, D. & Sternberg, L. D. L. (2000). The use of stable isotopes to study ecosystem gas exchange. *Oecologia, 123*, 297-311.

Yakir, D. & Wang, X. F. (1996). Fluxes of CO_2 and water between terrestrial vegetation and the atmosphere estimated from isotope measurements. *Nature, 380*, 515-517.

Yi, X. & Yang, Y. (2006). A stable carbon isotopic approach for understanding the CO_2 flux at the Haibei Alpine Meadow Ecosystem—A simple model. *Ecological Modelling, 193*, 796-800.

Zhang, F.-W., Liu, A.-H., Li, Y.-N., Zhao, L., Wang, Q.-X. & Du, M.-Y. (2008). CO_2 flux in alpine wetland ecosystem on the Qinghai-Tibetan Plateau. *Acta Ecologica Sinica, 28*, 453-462.

Zhang, J. X., Cao, G. M., Zhou, D. W., Hu, Q. W. & Zhao, X. Q. (2003). The carbon storage and carbon cycle among the atmosphere, soil, vegetation and animal in the *Kobresia humilis* alpine meadow ecosystem. *Acta Ecologica Sinica*, *23*, 627-634.

Zhao, L., Li, Y., Zhao, X., Xu, S., Tang, Y., Yu, G., Gu, S., Du, M. & Wang, Q., (2005). Comparative study of the net exchange of CO_2 in 3 types of vegetation ecosystems on the Qinghai-Tibetan Plateau. *Chinese Science Bulletin*, *50*, 1767-1774.

Zheng, D., Zhang, Q. & Wu, S. (2000). Mountain Geoecology and Sustainable Development of the Tibetan Plateau. Kluwer Academic, Dordrecht, the Netherlands.

Zhou, X. M. & Wu, Z. L. (2006). Vegetation and Plants Searchers List on The Haibei Research Station of Alpine Meadow Ecosystem, the Chinese Academy of Science. Qinghai People's Press.

In: Tundras: Vegetation, Wildlife...
Editors: B. Gutierrez et al. pp. 151-160
ISBN: 978-1-60876-588-1
© 2010 Nova Science Publishers, Inc.

Chapter 6

TUNDRAS AND CLIMATE CHANGE: A MAMMALIAN PERSPECTIVE

Věra Řičánková[1], Jan Robovský[1] and Petr Pokorný[2]

[1]Department of Zoology, Faculty of Science, University of South Bohemia, Branišovská 31, 370 05 Ceske Budejovice, Czech Republic.
[2]Institute of Archaeology, Academy of Sciences of the Czech Republic, Letenská 4, 118 01 Praha, Czech Republic.

ABSTRACT

Tundra ecosystems are seriously affected by global climate change. Understanding tundra history and postglacial development may enhance the ability of biologists to anticipate biotic responses to current environmental changes. Tundra-steppe was dominant vegetation type during Last Glacial and supported rich mammalian (mega)fauna.

Climate change represents critical factor responsible for the extinction of Glacial tundra-steppe fauna. Relatively constant climate in Central Asia allowed for a preservation of the tundra–steppe fauna in Altai-Sayan refugium. Arctic tundra could be considered rather a product of climate changes then anthropogenic activities and possibly represents a Holocene novelty, at least from a mammalian perspective. There is no evidence of extinction of arctic tundra species during Holocene. If the magnitude of current global warming anticipated by recent scenario remains unchanged, we may well expect the effect of climatic change analogous to that during "climatic optimum" of the Middle Holocene.

INTRODUCTION

Predicting how climate change will alter the future distributions of cold-loving animals and plants, and therefore the composition of arctic biota, is currently a major goal in ecology and conservation biology. The Arctic has experienced considerable warming in recent decades (an average of about 3°C) and climate projections suggest a continuation of the warming trend with an increase in mean annual temperatures of 4–5°C by 2080 (Callaghan et al., 2004a). Climate warming is expected to reduce the abundance and restrict ranges or even to cause the extinction (in specialists) of Arctic species rather than their evolving significantly in response to warming (Callaghan et al., 2004a). Migration of southerly taxa is very likely to occur more rapidly in the Artic than in other biomes. Latitudinal gradients suggest the higher sensitiveness of Arctic plant diversity to climate (Callaghan et al., 2004a). Degradation of permafrost (Anisimov & Reneva, 2006), displacement of tundra by forest, increase in the growth of vascular species with dominant woody plant over longer period (with generally disadvantaged mosses and liches), freeze-thaw cycles leading to ice-crust formation and consequent severe reduction of winter survival rate of a variety of animal species, deeper snow cover restricting access to winter pastures by not-subnival herbivores are also theoretically expected (Callaghan et al., 2004a). Recent warming is already impacting the Arctic ecosystems, disrupting population dynamics of Norwegian lemmings (Kausrud et al., 2008), reducing habitat area of polar bears (Durner et al., 2009) and affecting life history of migrating shorebirds (Piersma & Lindstrom, 2004). Rapid climate warming, however, is but the latest of many major climatic fluctuations to affect arctic biota during their history. Understanding tundra history and postglacial development may enhance the ability of biologists to anticipate biotic responses to current environmental changes.

PLEISTOCENE HISTORY

Tundra and tundra-steppe dominated a large area of northern Eurasia and North America at the Last Glacial Maximum approximately 15 000 years ago. Recent distinct landscape zones of steppes, forest-steppe, taiga, and tundra did not exist separately in the Last Glacial interval, but instead were altered to a single hyperzone in the most part of Northern Hemisphere (Guthrie, 2001; Vereshchagin & Baryshnikov, 1992).

Based on modern ecological observations and a certain amount of 'ecological extrapolation', it is possible to imagine a Glacial "tundra-steppe" as a mosaic of arctic-alpine and steppe species with patches of shrubs on damper sites. Trees have occurred in sheltered localities where there was adequate soil moisture and some shelter (Birks & Willis, 2008). Open-ground habitats of the Last Glacial consisted of a mosaics of highly productive continental loess steppe vegetation (*Cleistogenetea*), meadow-steppes with a mixture of alpine and xerotherm species in moister localities, slope- steppes on rocky substrate and relatively rare steppe-tundras *sensu stricto*. Steppe-tundras were found in moister areas with acid substrate and were characterized by low biomass production and high species richness (Jankovská et al., 2002; Kuneš et al., 2008).

Glacial vegetation was formed by cold and arid climate, pronounced seasonality and short vegetation seasons. Yet in this harsh environment, a conspicuously rich mammalian (mega)fauna prospered. The tundra-steppe supported a curious mix of species including woolly rhinos, bisons, lions, reindeer, horses, muskoxen, and mammoths (Guthrie, 1984, 1990; Tarasov et al., 1999). Snow cover represents a critical factor determining distribution of many recent species. During Last Glacial snow cover was very low, allowing herbivores to access their food resources during winter (Guthrie, 1990). Diversity of Late Pleistocene arctic mammalian communities was greater than present, animals reached larger body size and larger social organ size (antlers, horns) than their recent counterparts (Guthrie, 1984a,b; Musil, 1985; Geist, 1987).

PLEISTOCENE-HOLOCENE TRANSITION

This "mammoth steppe" community thrived approximately 100,000 years without major changes, and then suddenly went extinct about 10,000 years ago (see e.g. Martin & Klein, 1984; Guthrie, 2006). Global climate change associated with the beginning of the Holocene resulted in complete restructuring of Pleistocene landscapes.

The climate became more cloudy and humid, resulting in late thaw and a shallow active layer above the permafrost in summer. Waterlogged tussock tundra with a thick insulating moss layer developed over wide areas, preventing nutrient recycling (Guthrie, 2001). Grasses and arctic-alpine forbs were replaced by less productive mosses, lichens and dwarf birch shrubs at high latitudes (Guthrie, 1984b). Open tundra-steppe vegetation was replaced in lower latitudes by expanding forests. The most important characteristics of the Glacial vegetation – high productivity and heterogeneity – disappeared and tundra-steppe mosaic was replaced by relatively homogenous Holocene vegetation. High snow cover made winter grazing difficult or impossible and most of the shrub and tree species are toxic for herbivores (Guthrie, 1990).

As the result, many species of large herbivores went extinct and surviving arctic fauna became isolated from grasslands by taiga forests. The isolation enhanced further extinctions of smaller steppe species, including steppe lemmings or steppe pika (e.g. van Kolfschoten, 1995). The most important difference between recent and Pleistocene arctic mammalian fauna is the absence of species associated with grasslands, mainly large herbivores. Some tundra-steppe elements found refugia in alpine tundra of high mountain ranges e.g. Eurasian arctic hares, alpine marmots and chamois in Europe, Alpine or Siberian ibex in Eurasia (e. g. Musil, 1985; Wilson & Reeder, 2005).

ALTAI-SAYAN REFUGIUM

Whereas afforestation of much of western and central Europe began already in the last Late Glacial, when climatic conditions were relatively humid, farther east it did not occur until well into the Holocene because of much drier climate. This is consistent with the paleoclimatic model simulations that indicate higher summer temperatures related to higher insolation (Wright et al., 2003). These areas are too far inland to have been affected by the enhanced monsoonal rainfall that is found closer to the sources of atmospheric moisture. Extreme example of this are high mountain ranges in Southern Siberia. These Mountains (Altai and Sayan), represent a region where sharp climate fluctuations typical for Pleistocene-Holocene transition were not documented at all (Velichko et al., 1984; Chlachula, 2001). Recent findings of the paleo-biome reconstruction (Tarasov et al., 1998, 2000) and pollen-analytical research (Jankovská & Pokorný, 2008; Kuneš et al., 2008) suggest the present-day landscapes in some parts of southern Siberia and Mongolia are probably the closest modern analogy to the Last Glacial environments. This has been verified by comparison of recent pollen spectra from Altai–Sayan region and ones from the Last Full and Late Glacial of Central Europe (Kuneš et al., 2008; Pelánková et al., 2008). The detailed analysis of Altai Late Pleistocene assemblages of small mammals revealed there were no significant changes in species occurrence from the cold phase of Pleistocene to Holocene (Agadjanian & Serdyuk, 2005).

Recent Altai-Sayan mammalian fauna display some important characteristics typical for 'mammoth steppe' assemblages. Some Altai-Sayan mammalian communities represent a mixture of species of geographically distinct landscapes, whose distribution ranges usually do not overlap (e.g. Yudin *et al.*, 1979 and references therein). The south-eastern part of the region possibly represents the last remnant of the tundra-steppe with the typical Glacial faunal assemblages including horses, reindeer, moose, saiga, wolverine, pikas, Eurasian arctic hares, gray dwarf hamster, narrow-headed vole or steppe lemmings living sympatrically (Wilson & Reeder, 2005, Yudin et al., 1979). Very special position in the Altay-Sayan region has Mongolian saiga (*Saiga borealis*) a formerly widespread tundra-steppe species, surviving now just in the eastern part of the region (Baryshnikov & Tikhonov, 1994). Most of the recent steppe ungulates that inhabit Glacial refugium in the eastern Altai-Sayan region display similar size traits as their Pleistocene counterparts, being the largest of all Recent Palearctic subspecies. The Altai argali (*Ovis ammon ammon*), Mongolian gazelle (*Procapra gutturosa altaica*), and Altai wapiti (*Cervus canadensis sibiricus*) have absolutely and proportionally the largest horns or antlers of all Palearctic subspecies (Groves, 1967; Dolan, 1988; Heptner et al., 1988; Fedosenko & Blank, 2001, 2005). The largest size also exhibit local subspecies of wild horse tarpan (*Equus ferus przewalskii*) and kulan (dzigettai) (*Equus hemionus hemionus* and *E. h. castaneus*) (Groves 1974).

Thus, Altai-Sayan could be considered a refugium of Glacial ecosystems. Pleistocene fauna preserved especially in areas where forest belt does not separate alpine tundra from steppe grasslands. Alpine plants, together with steppe grasses were widespread during Last Glacial Maximum (Birks & Willis, 2008). Waterlogged tussock arctic tundra is unsuitable for alpine and steppe plants (Birks, 2008) as well as for herbivores depending on them. Guthrie (2001) suggests that recent arctic tundra may represent a Holocene novelty and existence of tundra-steppe refugium in Central Asia fully corroborate his opinion. Glacial tundra-steppe could be rather called "alpine-steppe".

CAUSES OF TUNDRA- STEPPE MAMMAL EXTINCTIONS

Holocene climate changes together with human exploitation are the most widely discussed factors responsible for extinction of the Glacial tundra- steppe fauna in the beginning of Holocene 10 000 BP (Martin & Klein, 1984; Zimov et al., 1995). Nevertheless, it is not clear whether a climate-induced change in vegetation caused extinction of the large herbivores (e.g. Guthrie, 2001) or whether loss of the herbivores from human hunting led to the change in vegetation (e.g. Zimov et al., 1995). Climate characteristics were probably more important factors determining extinction of large herbivores then anthropogenic factors, as megafauna extinction precede Holocene expansion of human settlement (Guthrie, 1984; 2006; FAUNMAP, 1996). Nonetheless, Zimov et al. (1995) suppose that introduction of North American buffalo herds in Yakutian tundra would change ecosystem functioning and restore Pleistocene tundra–steppe. Within Altai-Sayan refugium the tundra-steppe fauna, including large ungulates, persisted despite of human hunting pressure till at least the 16th century (Wilson & Reeder, 2005, Yudin et al., 1979; pers. observ.). Tundra-steppe fauna preserved in the region with relatively high human population densities and hunting pressure during Palaeolithic, but unaffected by Holocene climatic fluctuations. If the human hunting pressure were the main factor determining extinction or survival of large herbivores, then the Altai-Sayan populations would have been gone.

HOLOCENE DEVELOPMENT

Mammalian fauna occupying Holocene tundra experienced sharp climatic fluctuations in the last 10 000 years (Callaghan et al., 2004b). The most pronounced fluctuation occurred in Eurasia eight to six thousands years BP, during so called Holocene climatic optimum. Arctic ecosystems have been close to their minimum extent (Callaghan et al., 2004b). Surprisingly, the plant fossil record indicates that despite these major environmental changes, no arctic species has become extinct during the Quaternary, which emphasises the resilience of these plants to past environmental change (Birks, 2008). Mammalian fossil record from several arctic localities shows rather increase of species richness, as the boreal species advanced farther north and lived in sympatry with tundra species (Ukkonen, 1993; Boeskorov, 2006; Bachura & Kosintsev, 2006; Østbye et al., 2006). This example leaves good possibility that tundra species may survive recent global warming.

CONCLUSION

We conclude that climate change represents critical factor responsible for the extinction of Glacial tundra-steppe fauna. Relatively constant climate in Central Asia allowed for a preservation of the tundra–steppe fauna in Altai-Sayan refugium. Arctic tundra could be considered rather a product of climate changes then anthropogenic activities and possibly represents a Holocene novelty, at least from a mammalian perspective. If the magnitude of current global warming anticipated by recent scenario remains unchanged, we may well expect the effect of climatic change analogous to that during "climatic optimum" of the Middle Holocene. Holocene lesson shows that arctic plants and mammals have been able to cope with ecosystem changes induced by climate fluctuations.

ACKNOWLEDGMENTS

This work was supported by the Grant Agency of the Academy of Sciences of the Czech Republic grant #IAA6163303 and by Czech Ministry of Education grant #6007665801.

REFERENCES

Agadjanian, A. K. & Serdyuk, N. V. (2005). The history of mammalian communities and paleogeography of the Altai Mountains in the Paleolithic. *Paleontological Journal*, *39*, 645-821.

Anisimov, O. & Reneva, S. (2006). Permafrost and changing climate: the Russian perspective. *Ambio*, *35*, 169-175.

Bachura, O. & Kosintsev, P. (2007). Late Pleistocene and Holocene small- and large-mammal faunas from the Northern Urals. *Quaternary International*, *160*, 121-128.

Baryshnikov, G. & Tikhonov, A. (1994). Notes on skulls of Pleistocene saiga of Northern Eurasia. *Historical Biology*, *8*, 209-234.

Birks, H. H. (2008). The Late-Quaternary history of arctic and alpine plants. *Plant Ecology & Diversity*, *1*, 135-146.

Birks, H. J. & Willis, K. J. (2008). Alpines, trees, and refugia in Europe. *Plant Ecology & Diversity*, *1*, 147-160.

Boeskorov, G. G. (2006). Arctic Siberia: refuge of the Mammoth fauna in the Holocene. *Quaternary International*, *142*, 119-123.

Callaghan, T. V., Björn, L. O., Chernov, Y., Chapin, T., Christensen, T. R., Huntley, B., Ims, R. A., Johansson, M., Jolly, D., Jonasson, S., Matveyeva, N., Panikov, N., Oechel, W., Shaver, G., Elster, J., Henttonen, H., Laine, K., Taulavuori, K., Taulavuori, E. & Zöckler, C. (2004a). Climate Change and UV-B Impacts on Arctic Tundra and Polar Desert Ecosystems: Biodiversity, distributions and adaptations of Arctic species in the context of environmental change. *Ambio*, *33*, 404-417.

Callaghan, T. V., Björn, L. O., Chernov, Y., Chapin, T., Christensen, T. R., Huntley, B., Ims, R. A., Johansson, M., Jolly, D., Jonasson, S., Matveyeva, N., Panikov, N., Oechel, W., Shaver, G., Schaphoff, S., Sitch, S. & Zöckler, C. (2004b). Past changes in arctic terrestrial ecosystems, climate and UV radiation. *Ambio*, *33*, 398-403.

Chlachula, J. (2001). Pleistocene climate change, natural enviroments and palaeolithic occupation of the Altai area, west-central Siberia. *Quaternary International*, *80-81*, 131-167.

Dolan, J. M. (1988). A deer of many lands. A guide to the subspecies of the red deer *Cervus elaphus* L. *Zoonooz*, *62*, 4-34.

Durner, G. M., Douglas, D. C., Nielson, R. M., Amstrup, S. C., McDonald, T. L., Stirling, I., Mauritzen, M., Born, E. W., Wiig, Ø., DeWeaver, E., Serreze, M. C., Belikov, S. E., Holland, M. H., Maslanik, J., Aars, J., Bailey, D. A. & Derocher, A. E. (2009). Predicting 21st-century polar bear habitat distribution from global climate models. *Ecological Monographs*, *79*, 25-58.

FAUNMAP, Working Group (1996). Spatial response of mammals to Late Quaternary environmental fluctuations. *Science*, *272*, 1601-1606.

Fedosenko, A. K. & Blank, D. A., 2001. *Capra sibirica*. *Mammalian Species*, *675*, 1-13.

Fedosenko, A. K. & Blank, D. A., 2005. *Ovis ammon*. *Mammalian Species*, *773*, 1-15.

Geist, V. (1987). On the speciation in Ice Age mammals, with special reference to cervids and caprids. *Canadian Journal of Zoology*, *65*, 1067-1084.

Groves, C. P. (1967). On the gazelles of the genus *Procapra* Hodgson, 1846. *Zeitschrift für Säugetierkunde*, *32*, 144-149.

Groves, C. P. (1974). *Horses, asses and zebras in the wild*. London: David and Charles Newton Abbot.

Guthrie, R. D. (1984a). Alascan megabucks, megabulls, and megarams: the issue of Pleistocene gigantism. *Special Publication of Carnegie Museum of Natural History*, *8*, 482-510.

Guthrie, R. D. (1984b). Mosaics, allelochemics and nutrients. An ecological theory of Late Pleistocene megafaunal extinctions. In P. S. Martin & R. G. Martin (Eds.), *Quaternary Extinctions: A Prehistoric Revolution* (259-298). Tucson: The University of Arizona Press.

Guthrie, R. D. (2001). Origin and causes of the mammoth steppe: a story of cloud cover, woolly mammal. *Quaternary Review*, *20*, 549-574.

Guthrie, R. D. (2006). New carbon dates link climatic change with human colonization and Pleistocene extinctions. *Nature*, *441*, 207-209.

Heptner, V. G., Nasimovich, A. A. & Bannikov, A. G. (1988). *Mammals of the Soviet Union. Vol. I. Artiodactyla and Perissodactyla.* Washington D. C.: Smithsonian Institution Libraries and The National Science Foundation.

Jankovská, V. & Pokorný, P. (2008). Forest vegetation of the last full-glacial period in the Western Carpathians (Slovakia and Czech Republic). *Preslia, 80,* 307-324.

Jankovská, V., Chromý, P. & Nižnianská, M. (2002). Šafárka —first palaeobotanical data of the character of Last Glacial vegetation and landscape in the West Carpathians (Slovakia). *Acta Palaeobotanica, 42,* 39-50.

Kausrud, K. L., Mysterud, A., Steen, H., Vik, J. O., Østbye, E., Cazelles, B., Framstad, E., Eikeset, A. M., Mysterud, I., Solhøy, T. & Stenseth, N. Chr. (2008). Linking climate change to lemming cycles. *Nature, 456,* 93-987.

Kuneš, P., Pelánková, B., Chytrý, M., Jankovská, V., Pokorný, P. & Petr, L. (2008). Interpretation of the last-glacial vegetation of eastern-central Europe using modern analogues from southern Siberia. *Journal of Biogeography, 35,* 2223-2236.

Martin, P. S. & Klein, R. G. (1984). *Quaternary Extinctions: A Prehistoric Revolution.* Tucson: The University of Arizona Press.

Musil, R. (1985). Paleobiography of terrestrial communities in Europe during the Last glacial. *Acta Musei Nationalis Pragae, 41,* 1-84.

Østbye, E., Lauritzen, S. E., Moe, D. & Østbye, K. (2006).Vertebrate remains in Holocene limestone cave sediments: faunal succession in the Sirijorda Cave, northern Norway. *Boreas, 35,* 142-158.

Pelánková, B., Kuneš, P., Chytrý, M., Jankovská, V., Ermakov, N., Svobodová-Svitavská, H. (2008). The relationships of modern pollen spectra to vegetation and climate along steppe-forest-tundra transition in southern Siberia, explored by decision trees. *The Holocene, 18,*1259-1271.

Piersma, T. & Lindström, Å. (2004). Migrating shorebirds as integrative sentinels of global environmental change. *Ibis, 146,* S61-S69.

Tarasov, P. E., Webb, T., Andreev, A. A., Afanas'eva, N. B., Berezina, N. A., Bezusko, L. G., Blyakharchuk, T. A., Bolikhovskaya, N. S., Cheddadim, R., Chernavskaya, M. M., Chernova, G. M., Dorofeyuk, N. I., Dirksen, V. G., Elina, G. A., Filimonova, L. V., Glebov, F. Z., Guiot, J., Gunova, V. S., Harrison, S. P., Jolly, D., Khomutova, V. I., Kvavadze, E. V., Osipova, I. M., Panova, N. K., Prentice, I. C., Saarse, L., Sevastyanov, D. V., Volkova, V. S. & Zernitskaya, V. P. (1998). Present-day and mid-Holocene. *Journal of Biogeography, 25,* 1029-1053.

Tarasov, P. E., Peyron, O., Guiot, J., Brewer, S., Volkova, V. S., Bezusko, L. G., Dorofeyuk, N. I., Kvavadze, E. V., Osipova, I. M. & Panova, N. K. (1999). Last Glacial Maximum. *Climate Dynamics*, *15*, 227-240.

Tarasov, P. E., Volkova, V. S., Webb, T., Guiot, J., Andreev, A. A., Bezusko, L. G., Bezusko, T. V., Bykova, G. V., Dorofeyuk, N. I., Kvavadze, E. V., Osipova, I. M., Panova, N. K. & Sevastyanov, D. V. (2000). Last Glacial maximum biomes reconstructed from pollen and plant macrofossil data from northern Eurasia. *Journal of Biogeography*, *3*, 609-620.

Ukkonen, P. (1993). The Postglacial History of the Finnish Mammalian Fauna. *Annales Zoologici Fennici*, *30*, 249-264.

Van Kolfschoten, T. (1995). On the application of fossil mammals to the reconstruction of the palaeoenviroment of northwestern Europe. *Acta Zoologica Cracoviensia*, *38*, 73-84.

Velichko, A. A., Wright, H. E. &. Barnosky, C.W. (1984). *Late Quaternary enviroments of the Soviet Union*. London: Logman.

Vereshchagin, N. K. & Baryshnikov, G. F. (1992). The ecological structure of the "Mammoth Fauna" in Eurasia. *Annales Zoologici Fennici*, *28*, 253-259.

Wilson, D. E. & Reeder, D.-A. M. (2005). *Mammal species of the World. A taxonomic and geographic reference*, 3rd edition. Baltimore: The Johns Hopkins University Press.

Wright, H. E. Jr., Ammann, B., Stefanova, I., Atanassova, J., Margalitadze, N., Wick, L., Blyakharchuk, T. (2003). Late-glacial and Early-Holocene Dry Climates from the balkan Peninsula to Southern Siberia. In S.Tonkov (Ed.), *Aspect of palynology and Palaeoecology. Festshrigt in honour of Elisaveta Bozilova* (127-136). Sofia – Moscow: Pensoft Publishers. 127-136.

Yudin, B. S., Galkina, L. I. & Potapkina, A. F. (1979). *Mlekopitayushchie Altae-Sayanskoi gornoi strany* [Mammal s of the Altai-Sayan mountainous part]. Novosibirsk: Nauka.

Zimov, S. A., Chuprynin, V. I., Oreshko, A. P., Chapin III, F.S., Reynolds, J. F. J. & Chapin, M.C. (1995). Steppe-tundra transition: a herbivore-driven biome shift at the end of the Pleistocene. *American Naturalist*, *146*, 765-794.

In: Tundras: Vegetation, Wildlife...
Editors: B. Gutierrez et al. pp. 161-173

ISBN: 978-1-60876-588-1
© 2010 Nova Science Publishers, Inc.

Chapter 7

CARBON DEPOSITION ON THE FORESTS OF SOME TREELINE ECOTONES OF THE URAL FEDERAL DISTRICT

A. Usoltsev Vladimir

Botanical Garden of Russian Academy of Sciences, Ural Branch,
8 Marta Str., 202, Ekaterinburg, 620144, Russia

ABSTRACT

Carbon deposition on the forests of two treeline ecotones is studied. The first of them is on the altitudinal gradient of the western slope of Tylaiskii Kamen Mountain (the western part of Konzhakovskii-Tylaiskii-Serebryanskii Mountain system, 59^0 30' N, 59^0 00' E), namely on the belt of the upper tree-limit rising between 864 and 960 m above see level, during the last 100 years. The second one is a zonal ecotone near the lower Pur river (67^0 N, 78^0 E) as the transition belt between closed flood-lands forests and island-like forests on the tundra watershed varying from tens of meters to 2-3 kilometers depending of relief peculiarities and flood-lands width. On the first treeline ecotone the 5-6 times decreasing of the carbon pool in *Picea* biomass between altitudinal levels of 864 and 960 m a.s.l. was recognized. On the second ecotone at the age of 45 years and similar densities (1300-1700 trees per ha) the carbon pool in *Larix* aboveground biomass and needle biomass on the flood-lands are correspondingly 7,0 and 2,4 times more than on the watershed. In senescent forests this difference is some more, correspondingly 10 and 3 times. Annual carbon deposition differs 5 times by these two sites.

Keywords: Treeline ecotone, carbon pool and deposition, Tylaiskii Kamen Mountain, the lower Pur river.

Treeline ecotone as a belt of transition from forest vegetation to nonforest one (tundra, steppe, swamp, etc.) is an unique natural object allowing to monitor climate change consequences (Kullman, 1990; Korner, 1999; Bugmann, Pfister, 2000; Holtmeier, 2003; Shiyatov, 1995, 2003).

THE OBJECT STUDIED

Our studying forest biological productivity on the altitudinal treeline ecotone is carried out on the western slope of Tylaiskii Kamen Mountain (the western part of Konzhakovskii-Tylaiskii-Serebryanskii Mountain System, 59^0 30' N, 59^0 00' E), namely on the belt of the upper tree-limit rising between 864 and 960 m above see level, during the last 100 years (Figure 1 and 2).

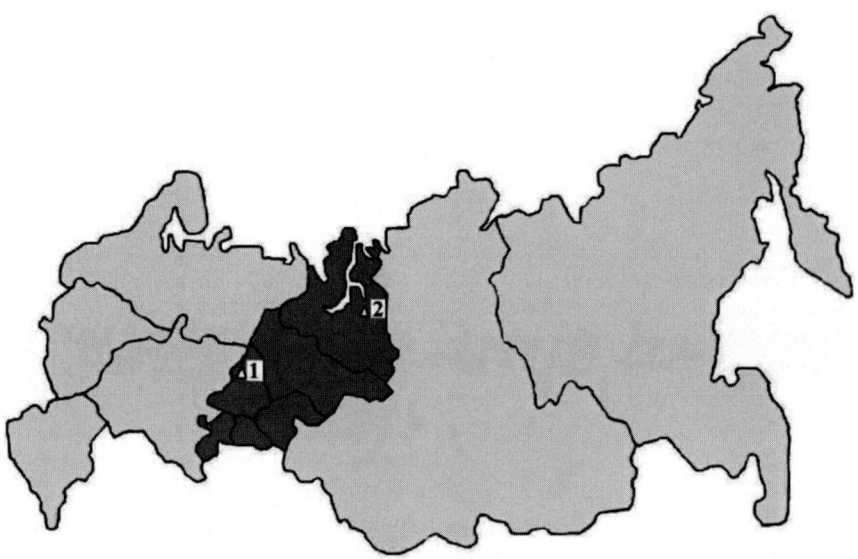

Figure 1. Sample plot locations on the altitudinal (1) and zonal (2) treeline ecotones forest-tundra on Konzhakovski Kamen Mountains and the lower Pur river correspondingly on the area of the Ural Federal District.

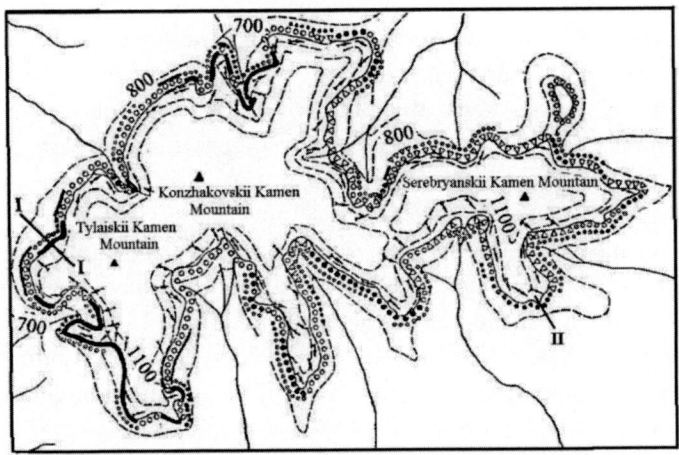

Figure 2. The location of the profile transect (I-I) on the ecotone forest-tundra on the western slope of Tylaiskii Kamen Mountain (as a part of Konzhakovskii-Tylaiskii-Serebryanskii mountain system) where sample plots were established. The upper timberline (II) composed by *Picea obovata, Pinus sibirica, Larix Sukaczevii* and *Betula tortuosa* was mapped by P.L. Gorchakovskii and S. G. Shiyatov (1970). The numbers are heights above see level.

On the zonal ecotone near the lower Pur river (67^0 N, 78^0 E) the transition belt width between closed flood-lands forests and island-like forests on the tundra watershed varies from tens of meters to 2-3 kilometers depending of relief peculiarities and flood-lands width. The shore drainage, silt sediments and warmth (thermal) river flowing are the causes of high *Larix* forest yield on the riverside belt of forest-tundra, the same as on taiga zone. When one is moving away from the flood-lands, *Larix* forest site index is decreased from II-III (by Orlov scale) on the lower terrace to V-Va on the elevated sites of the watershed.

THE METHODS USED

The bulk of the model trees was studied on the temporary sample plots established from July to August. The methods used were different in the cases biomass and primary production estimating.

Biomass estimating. The stem biomass was determined by measuring stem diameters outside and inside bark at 10 cross sections with subsequent conversion of the volume indices to weight indices with the use of the data on the basic density calculated for each of 10 discs sawn out. The biomass of the crown was divided into three equal parts, each of them was weighed completely and then the proportion of foliage was determined either by direct picking off the leaves (for small trees) or by cutting off the the woody greens (branch parts covered with foliage) by pruning shears (for larger trees). The variation of the proportion of foliage in the mass of the woody greens is only 2-5 % (Usoltsev, 1988), it was determined for model branches taken over the profile of the crown. To convert the fresh mass of branches to a dry state, we took from each model tree, regardless of position in the crown, three discs – from large, middle and small branches.

Root biomass was determined by the method by A.F. Tchmyr (1984). It suggested the two stages. On the first of them the stump monolite is digged by spade and the proximal roots are exposed on the area of 30-100 cm from the stump to the depth of penetration of tap root. On the second stage the small monolites (25×20×10 cm) in a number of about 10 are established along proximal roots. The root mass obtained for these stages is extrapolated on the area of stump and proximal root expansion and then – on the area of tree growth. Thin roots on the area between proximal roots were not determined.

Primary production estimating (only on lower Pur river, *Larix* forests). The stem primary production was determined by measuring stem diameters inside bark on this moment and 5 years ago at 10 cross sections with subsequent conversion of the volume 5-year difference to weight indices with the use of the data on the basic density calculated for each of 10 discs sawn out. Primary production of *Larix* foliage supposed to be equal to its biomass determined not only on branches but also on the stems. Primary production of tree branches is determined using the method by A.I. Rusalenko and E.G. Petrov (1975) by means of dividing double branch mass of the whole tree by the crown age measured on stem rings just below crown. This method supposes the fact that the branch of mean mass is in the middle point of the crown (Utkin, 1975). Our special studying (Usoltsev and Zalesov, 2005) shows this method is more exact than the method by R.H. Whittaker (1965). Root primary production was not determined.

RESULTS AND DISCUSSION

Altitudinal Treeline Ecotone

It was recognized that about 100 years ago on the lower level (864 m a.s.l.) a mixed spruce-fir-birch forest was established. Approximately at the same time a more high position up-slope (on the middle level 924 m a.s.l.) was occupied by *Abies sibirica* trees too. About 40 years later the upper level (960 m a.s.l.) was occupied by *Pinus sibirica* trees and some more after 12 years the same level was reached by *Picea obovata* trees.

Because of progressive rising the upper tree limit, the *Picea* forest age is decreased from 100 years at the lower level to 28 years at the upper level. Thus, following to specific temporal dynamics of environment, at different time periods different species were as pioneers rising up-slope.

The 5-6 times decreasing of the carbon pool in *Picea* biomass between altitudinal levels of 864 and 960 m a.s.l. was recognized. The total decreasing for all the species was 9 time correspondingly (Table 1, Figure 3).

If one supposing forest biomass mortality and soil respiration are balanced (Shvidenko et al., 2001), average carbon pool change is equal to annual net ecosystem production (NEP) (t C per ha) for the period of development of trees on the given altitudinal level. With calculating NEP for each level, decreasing of annual NEP up-slope from 1,60 (864 m a.s.l.) to 0,67 t C per ha (960 m a.s.l.) on the belt of upper tree-limit rising was recognized (Table 2).

On the contrary, the share of root carbon in the total pool is increased up-slope: *root carbon: aboveground carbon* ratio is 0,07; 0,21 and 0,24 and *root carbon: stem carbon* ratio is 0,15; 0,32 and 0,58 on the corresponding levels of 864, 924 and 960 m a.s.l.

Zonal Ecotone Near the Lower Pur River.

At the age of 45 years and similar densities (1300-1700 trees per ha) the carbon pool in *Larix* aboveground biomass and needle biomass on the flood-lands are correspondingly 7,0 and 2,4 times more than on the watershed. In senescent forests this difference is some more, correspondingly 10 and 3 times. Annual carbon deposition differs 5 times by these two sites (Figure 4).

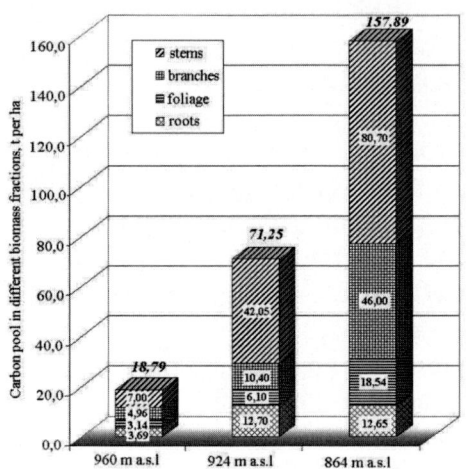

Figure 3. Summary carbon pools in total biomass of the forest-forming species according to the altitudinal gradient of Tylaiskii Kamen Mountain.

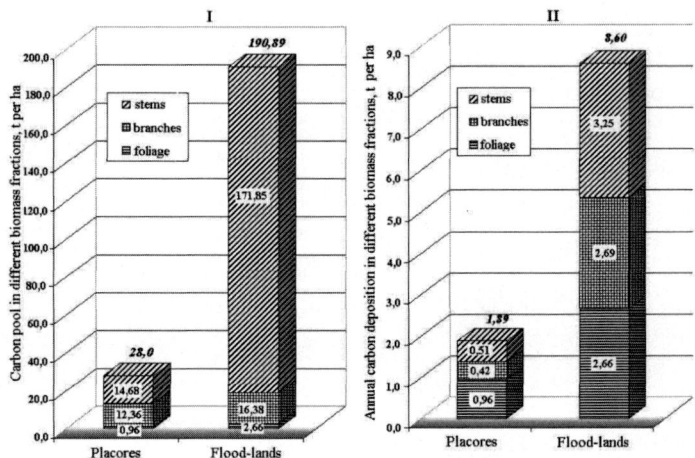

Figure 4. Carbon pools (I) and their annual depositions (II) in biomass of pure larch forests on the zonal treeline ecotone forest-tundra at the lower Pur river.

Nevertheless both flood-lands forests and elevated ones have similar relative carbon fraction composition in annual aboveground carbon: the shares of stems, needles and branches on the first case are correspondingly as 38, 40 and 22 % and on the second case – 30, 40 and 30 % (Table 3).

Table 1. Carbon pool in forest biomass of the basic forest-forming species according to the altitudinal gradient of Tylaiskii Kamen Mountain on the belt of the upper tree-limit rising between 864 and 960 m above see level, during the last 100 years.

Tree species	A, yrs	H, m	D, cm	G, m²/ha	N, trees/ha	Organic carbon in forest biomass, t/ha						
						Cs	Csb	Cf	Cb	Cabo	Cr	Ctot
Altitudinal level 1 (960 m a.s.l.)												
Picea obovata	28	2,0	7,0	18,4	4800	6,25	0,60	3,12	4,90	14,87	3,65	18,52
Pinus sibirica	40	1,4	4,0	0,5	425	0,12	0,03	0,02	0,06	0,23	0,04	0,27
Total	30	-	-	18,9	5225	6,37	0,63	3,14	4,96	15,10	3,69	18,79
Altitudinal level 2 (924 m a.s.l.)												
Picea obovata	79	3,3	15,9	54,5	2750	35,10	1,75	5,34	9,10	51,29	11,10	62,39
Abies sibirica	102	2,2	8,8	7,7	1250	4,95	0,25	0,76	1,30	7,26	1,60	8,86
Total	80	-	-	62,2	4000	40,05	2,00	6,10	10,40	58,55	12,70	71,25
Altitudinal level 3 (864 m a.s.l.)												
Picea obovata	99	7,2	22,7	65,8	1625	41,40	2,70	12,30	27,20	83,60	6,25	89,85
Abies sibirica	101	3,3	7,7	18,1	725	16,65	1,05	0,85	7,15	25,70	3,75	29,45
Betula alba	97	7,0	23,1	28,2	1825	17,75	1,15	5,39	11,65	35,94	2,65	38,59
Total	100	-	-	112,1	4175	75,80	4,90	18,54	46,00	145,24	12,65	157,89

Table 2. Net ecosystem production (NEP) change along altitudinal gradient on the treeline ecotone forest-tundra of Tylaiskii Kamen Mountain

Tree species	Tree age, yrs	Carbon pool in total biomass, t/ha	Annual NEP, t/ha
Altitudinal level 1 (960 m a.s.l.)			
Picea obovata	28	18,52	0,66
Pinus sibirica	40	0,27	0,01
Total	30	18,79	0,67
Altitudinal level 2 (924 m a.s.l.)			
Picea obovata	79	62,39	0,79
Abies sibirica	102	8,86	0,09
Total	80	71,25	0,88
Altitudinal level 3 (864 m a.s.l.)			
Picea obovata	99	89,85	0,91
Abies sibirica	101	29,45	0,29
Betula alba	97	38,59	0,40
Total	100	157,89	1,60

Designations: A – tree average age, years; H – tree mean height, m; D – tree mean diameter at breast height, cm; G – basal area, m^2/ha; N – tree number per ha; Cs, Csb, Cf, Cb, $Cabo$, Cr, $Ctot$ are correspondingly carbon pool in biomass of stems above bark, stem bark, foliage, branches, aboveground biomass, roots and total biomass, t per ha.

If we try to compare the average carbon pools in biomass of *Larix sibirica* in the flood-lands of the Pur river with similar indices of *L. gmelinii* on Estern Siberia on the same latitudes (Pozdnyakov, 1975; Mitrofanov, 1984; Shepashenko et al., 2001), we can find the almost 5-fold exceeding the first of them above the second (Table 4). A cause of this matter may be much more climate continentality or an other related factor (for example, precipitation or the sum of positive temperatures) on Eastern Siberia than near the Pur river.

Nevertheless on placores the average carbon pool in biomass of *L. sibirica* (lower Pur river, Western Siberia), *L. gmelinii* (Putorana Plateau, Middle Siberia) and *L. cajanderi* (Yana river basin, Eastern Siberia) are similar and are correspondingly 8,0; 9,7 и 9,3 t per ha (Table 4). It seems to be the factor limiting larch growth at all these sites the same one, namely long-term permafrost.

Table 3. Carbon pool and its annual deposition in larch biomass on treeline ecotone forest-tundra of the lower Pur river

Taxation indices							Carbon of biomass, t per ha					Annual carbon deposition, t per ha				
A, yrs	Site index	D, cm	H, m	G, m²/ha	N, trees/ha	Stem volume, m³/ha	Stems total	Stem bark	Bran-ches	Foli-age	Total	Stems total	Stem bark	Bran-ches	Foli-age	Total
High elevations (placores), vacciniosum tipe																
45	V	6,6	7,6	5,96	1740	24,2	4,79	0,97	1,28	0,33	7,36	0,34	0,08	0,10	0,33	0,84
100	Va	11,9	9,5	7,57	677	38,0	5,88	2,42	1,41	0,48	10,2	0,06	0,02	0,12	0,48	0,67
102	Va	10,9	9,3	5,16	550	25,2	4,01	1,54	0,67	0,15	6,36	0,11	0,05	0,20	0,15	0,50
Flood-lands, hylocomiosum type																
45	II	15,0	15,2	23,6	1329	200	42,45	6,20	4,55	0,81	54,00	2,01	0,25	0,93	0,81	3,98
260	III	31,3	23,7	40,8	944	446	84,05	19,90	7,05	1,27	112,3	0,57	0,13	0,27	1,27	2,23
350	II	24,0	31,5	21,7	484	218	45,35	7,75	4,78	0,58	58,46	0,67	0,14	0,49	0,58	1,87

Table 4. The carbon pool values in above-ground biomass of tundra larch forests on Middle and Eastern Siberia.

№	Forest type	Site index	A, yrs	Carbon pool in biomass, t/ha					Reference
				Stems total	Stem bark	Branches	Foli-age	Total	
Middle Siberia, Taimyr Peninsula, locality "Ary-Mas". 72°28'N, 101° E. Tundra, *Larix gmelinii*									
1	Led.	Vc	142	3,44	-	1,36	0,22	5,02	Knorre, 1977
2	Aln.	Vc	142	2,07	-	1,00	0,09	3,16	
3	Car.	Vc	142	1,62	-	1,25	0,11	2,98	
4	Car.	Vc	142	0,15	-	0,21	0,03	0,39	
5	Car.	Vc	142	0,15	-	0,09	0,013	0,25	
Middle Siberia, Putorana Plateau. 70°N, 90° E Forest-tundra, *Larix gmelinii*									
1	Aln.	V	155	25,6	-	1,53	0,55	27,7	Deyeva, 1985,1987
2	Vac.	Vc	155	1,51	-	0,20	0,045	1,75	
3	Frut.	V	150	6,60	-	2,75	0,23	9,58	Pautova, 1976
4	Aln.	V	150	7,65	-	3,10	0,22	11,0	
5	Aln.	V	150	7,95	-	3,25	0,27	11,5	
6	Led.	V	150	12,4	-	3,90	0,31	16,6	
7	Aln.	V	150	14,4	-	3,30	0,36	18,1	
8	Frut.	Vc	150	1,10	-	0,50	0,07	1,67	

Table 4. (Continued)

Eastern Siberia, Jakutiya, Zhigansk, Verkhoyansk. 67^0 N, 123^0 E. Forest-tundra, placores, *Larix cajanderi*								
1	*Led.*	Vb	300	14,9	2,85	1,90	0,27	17,1	Pozdnya-kov, 1975; Митрофа-нов, 1984
2	*Led.*	Va	190	18,9	3,65	1,00	0,50	20,4	
3	*Frut.*	Va	85	5,1	1,00	0,60	0,27	5,97	
4	*Led.*	V6	350	15,7	2,45	1,35	0,31	17,4	
5	*Frut.*	Va	150	15,5	2,80	1,05	0,18	16,7	
6	*Vac.*	V	170	24,7	4,10	1,10	0,36	26,2	
Eastern Siberia, Jakutiya, Zhigansk, Verkhoyansk. 67^0 N, 123^0 E. Forest-tundra, valley, *Larix cajanderi*								
1	*Ulig.*	IV	60	21,5	4,15	3,65	0,54	25,7	Mitrofa-nov, 1984
2	*Vac.*	IV	255	21,9	3,35	1,15	0,50	23,6	
3	-	V	90	3,05	0,60	0,45	0,23	3,73	
4	*Frut.*	IV	32	7,45	1,65	1,05	0,58	9,08	
5	-	V	44	14,6	3,25	0,55	0,36	15,5	
6	*Vac.*	Va	200	12,9	2,50	1,75	0,72	15,4	
Eastern Siberia, Jakutiya, Djanky river mouth. 67^0 N, 133^0 E. Forest-tundra, placores, *Larix cajanderi*								
1	*Sphag.*	Vc	113	6,27	1,25	0,79	0,19	7,25	Shepa-shenko et al., 2001
2	*Sphag.*	Vb	124	14,4	2,84	1,92	0,45	16,8	
3	*Sphag.*	Vc	115	3,38	0,68	0,48	0,12	3,98	
Forest-tundra, flood-lands, *Larix cajanderi*								
1	*Vac.*	Va	125	20,0	3,86	6,41	0,66	27,1	
2	*Ulig.*	Vb	127	8,58	1,73	1,08	0,27	9,93	
3	*Car.*	Vc	92	1,62	0,34	0,13	0,04	1,79	

ACKNOWLEDGMENTS

We thank Dr. Galako V.A. and Bogoslovskaya O.A. for their kind cooperation. This research is supported in part by the Russian Foundation for Fundamental Research under the grants «Regional regularities of carbon depositing in forest ecosystems of Russia» (No. 00-05-64532) and «Estimating the carbon pool and the carbon-oxigen budget of forest ecosystems of the Urals region» (No. 01-04-96424), as well as by the project INTAS 2001-0052.

REFERENCES

Bugmann, H. & Pfister, C. (2000). Impacts of interannual climate variability on past and future forest composition. *Reg. Environ. Change*, *1*, 112-125.

Deyeva, N. M. (1985). Biomass amount in forest communities of the northern-western part of Plateau Putorana. *Rus. Botanical Journal*, *70*, 54-58 (in Russian).

Deyeva, N. M. (1987). Vegetative mass structure in forest phytocoenoses of the northern-western part of Plateau Putorana. *Rus. Botanical Journal*, *72*, 505-511 (in Russian).

Gorchakovskii, P. L. & Shiyatov, S. G. (1970). Physionomic and ecological differentiation of upper timberline on the Northern Ural. Misc. Sverdlovsk Branch of the All-Union Botanical Society. *Sverdlovsk, issue* 5, 14-33 (in Rus.).

Holtmeier, F.-K. (2003). Mountain Timberlines. *Ecology, Patchiness, and Dynamics*.– Dordrecht, Boston, London: Kluwer Academic Publishers. 369 p.

Knorre, A. V. (1977). Larch aboveground biomass in basic communities of the "Ary-Mas" forest. Transactions of State Reserve "Stolby". *Krasnoyarsk, Issue*, *11*, 91-100 (in Russian).

Korner, Ch. (1999). Alpine Plant Life. Berlin, Heidelberg: *Springer-Verlag*, 343.

Kullman, L. (1990). Dynamics of altitudinal tree-limits in Sweden: a review. *Nor. Geogr. Tidsskr*, *44*, 103-116.

Mitrofanov, D. P. (1984). Estimating Productivity of the Siberian northern-taiga forests. *In: Productivity of forest phytocoenoses*. Krasnoyarsk: Sukachev Forest Institute. 95-102. (in Russian).

Pautova, V. N. (1976). Aboveground plant mass and transpiration of some communities in forest-tundra zone. *Transactions of Limnological Inst.*, *22(42)*, 92-128 (in Russian).

Pozdnyakov, L. K. (1975). Productivity of Siberian Forests. *In: Biospheric resources: Synthesis of the Soviet studies for the International Biological Programme. Vol. 1.* Leningrad: Nauka, 43-55 (in Russian).

Rusalenko, A. I. & Petrov E. G. (1975). Determining biomass increment in pine forests. *In: The current growth of forest stands.* Minsk, Uradzhai Publ. P. 139-140 (in Russian).

Shiyatov, S. G. (1995). Reconstruction of climate and upper timberline dynamics since AD 745 by tree-ring data in the Polar Ural Mountains. In: International conference on Past, Present and Future Climate / Ed. Henkinheimo Pirkko. Painatuskeskus: Publication of Academy of Finland, 6/95: 144 – 147.

Shiyatov, S. G. (2003). Rates of change in the upper treeline ecotone in Polar Ural Mountains. *PAGES News, 11, 1,* 8-10.

Shvidenko, A. Z., Nilsson, S., Stolbovoi, V. S., Rozhkov, V. A. & Gluck, M. (2001). Experience of aggregated estimation of basic indices of bioproductive process and carbon budget at terrestrial ecosystems of Russia. 2. Net primary production of ecosystems. *Rus. J. Ecology, 2,*83-90.

Tchmyr, A. F. (1984). Forest plantations: methodical guidelines on tree root system investigating. *Leningrad, Forest Engineering Academy,* 40 (in Russian).

Usoltsev, V. A. (1988). Principles and methods of compiling stand bioproductivity tables. *Soviet Forest Sciences, 2,* 23-32 (Allerton Press Inc.).

Usoltsev, V. A. & Zalesov S. V. (2005). *Methods of determining biological productivity of forests.* Ekaterinburg, Ural State Engineering Academy, 147 p. (in Russian).

Utkin, A. I. (1975). Biological productivity of forests: methods of studying and results. *In: Forestry and forest management: Science and engineering totals.* Moscow: VINITI. *Vol. 1,* 9-189 (in Russian).

Whittaker, R. H. (1965). Branch dimension and estimation of branch production. *Ecology, 46,* 365-370.

In: Tundras: Vegetation, Wildlife...
Editors: B. Gutierrez et al. pp. 175-192

ISBN: 978-1-60876-588-1
© 2010 Nova Science Publishers, Inc.

Chapter 8

GENETIC DIVERSITY AND POPULATION STRUCTURE OF ALPINE PLANTS ENDEMIC TO QINGHAI-TIBETAN PLATEAU, WITH IMPLICATIONS FOR CONSERVATION UNDER GLOBAL WARMING

Yupeng Geng[1], John Cram[2] and Yang Zhong[3]
[1]School of Life Sciences, Fudan University, Shanghai 200433, China.
[2]China-UK HUST-Rres Genetic Engineering and Genomics Joint Laboratory,
Huazhong University of Science and Technology,
Wuhan 430074, China.
[3]School of Life Sciences, Fudan University, Shanghai 200433,
and Institute of Biodiversity Science and Geobiology,
Tibet University, Lhasa 850000, China.

ABSTRACT

The Qinghai-Tibetan Plateau is one of the most important centers of biodiversity for alpine species in the world and is among the areas that are most sensitive to global warming. Knowledge about population genetics is essential for understanding the dispersal ability and evolutionary potential of alpine species in a warming world. In this chapter, we review the genetic diversity and population structure of 19 alpine plant species endemic to the Qinghai-Tibetan Plateau. Generally, the population genetic variation can varygreatly among different species and the endangered species have much lower levels of genetic diversity than the co-occurring common species. Although a few species showed increased levels of genetic diversity along altitude, we dectected no significant correlation between diversity and altitude in most species. In addition, the isolation-by-distance model cannot explain the spatial genetic structure in most alpine species that have been investigated, which may partially due to the discontinous distribution of alpine species shaped by complex geomorphology in Qinghai-Tibetan Plateau. The implications of these results for the conservation of alpine plants during global warming are discussed.

INTRODUCTION

The Qinghai-Tibetan Plateau is the highest and largest plateau in the world. The major part of the plateau is located in China and has an area of 2.5 million km2, including Qianghai Province, the Tibet Autonomous Region and part of adjacent Chinese provinces (e.g., Gansu, Sichuan, and Yunnan, as well as Xinjiang Autonomous Region, see Figure 1). As a geographic term, the plateau also includes a few adjacent regions (N 25°-40°, E 74°-104°) outside China, in Nepal, Bhutan, Afghanistan, Pakistan, Tajikistan, Kyrgyzstan, and India.

The Qinghai-Tibetan Plateau is often called the 'third polar region' of the earth, comparable to the Arctic and the Antarctic. With an average altitude of more than 4000 m, it is a unique biogeographic region, where various landscapes, altitudinal belts, alpine ecosystems, and endangered and endemic species have developed. In particular, the southeastern part of Qinghai-Tibetan Plateau belongs to one of the 25 global biodiversity hotspots ("the South-Central China area") (Myers et al. 2000). Taking vascular plants as a sample, it is estimated that the plateau contains more than 12,000 species in 1500 genera, among which more than 20% are endemic to this region (Wu 1988; Wu et al. 1995). Additionally, many wild relatives of crops (e.g. wild barley) and medical plants (e.g. *Rhodiola* species) are distributed in the plateau. The rich biodiversity and unique environments make the Qinghai-Tibetan Plateau a special laboratory for botanists, ecologists, and evolutionary biologists.

The plateau is also among the areas of the world that are most sensitive to global warming (Qin 1998; Weng and Zhou 2006; Xu and Liu 2007). A warmer climate and altered precipitation patterns may have significant effects on the composition and distribution of biodiversity in alpine areas (Baker and Moseley 2007). In addition to the changing climate, intensive anthropic disturbance (e.g. overgrazing by livestock or over-harvesting of medical plants) will further accelerate habitat loss and environmental degration in the plateau. As a result, biodiversity in the Qinghai-Tibetan Plateau is at great risk and effective conservation efforts are urgently needed. Knowledge of genetic diversity and population structure provide important bases for the conservation of threatened species, as the long-term survival of a species largely depends on the maintenance of genetic variability within and among populations to accommodate new selection pressures resulting from inevitable environmental changes like global warming (Kinnison et al. 2007).

In this chapter, the general levels and patterns of genetic variation for plant species endemic to the Qinghai-Tibetan Plateau are outlined. We focus on intraspecific genetic variation and microevolution processes, in an attempt to provide a better understanding of the potential ecological and evolutionary responses of alpine plants in a warmer world. Recently, the macroevolution patterns and processes in the Tibetan flora (e.g. adaptive radiation and phylogenetic relationship between closely related species) have received increased attention. This topic is interesting but beyond the scope of this review. Furthermore, we have not attempted to make a comprehensive review of all published work, but have instead paid more attention to papers published in recent years. Our aim is to assess the current state of this field and raise a few questions that deserve attention in future work.

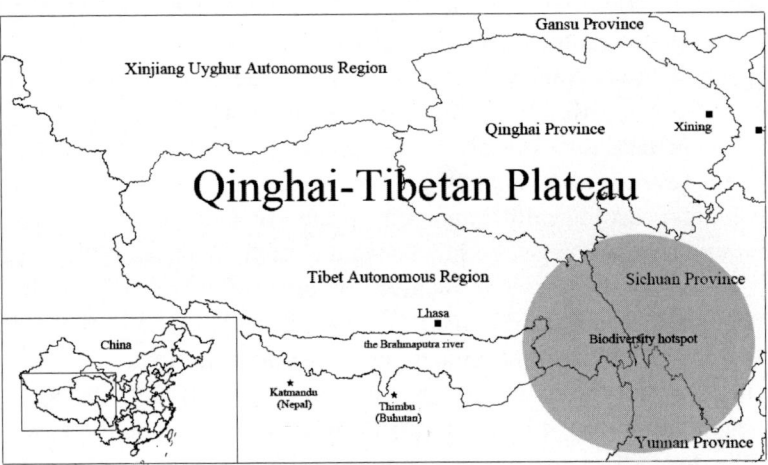

Figure 1. Location and range of the Qinghai-Tibetan Plateau. The shaded area indicates the approximate range of one of the 25 global biodiversity hotspots - South-Central China.

BRIEF HISTORY OF PLANT GENETIC DIVERSITY STUDIES IN QINGHAI-TIBETAN PLATEAU

The earliest exploration of the Qinghai-Tibetan Plateau by western explorers goes back to 19th century, started with the plant collection from Sikkim to Tibet by Joseph Dalton Hooker (Liu 2000). In the 1930s-40s, several Chinese scientists, e.g. Shen'e Liu (T. N. Liou), Dejun Yu (T. T. Yu), Jinzhi Xu, and JianChu Sun, also conducted a few scientific investigations of botany, geography, and meteorology. But these activities were confined to limited areas until the foundation of the Peoples Republic of China in 1949. The systematic scientific investigation of the Qinghai-Tibetan Plateau began in the 1950s. A large group was set up by the Chinese Academy of Sciences (also called "Academia Sinica") in 1973 and conducted extensive investigations during the following twenty years (Liu 2000). Based on large-scale collections of plant specimens, a number of monographs including the *Flora of Tibet* and *Vegetation of Tibet* have been published, which are the most important basic data sources for studies of the plants in the Qinghai-Tibetan Plateau even today.

In the early stages of investiating plant genetic diversity in Tibet, many studies focused on the collection and evaluation of germplasm resources in crops, especially barley (*Hordeum vulgare* L. var. *nudum* Hook. f.), one of the major crops in the plateau, and their wild relatives. Zhou et al. (1984) analyzed the karyotype and chromosome Giemsa N-banding patterns of two-rowed wild barley (*H. spontaneum* C. Koch). Using isozyme markers, Dai and Zhang (1989) analyzed the genetic diversity of 463 accessions of cultivated barley. Recently, in addition to several economically important species (e.g. crop relatives and medical plants), ecologically important species (e.g. dominant species in local communites) have also received increasing attention. With regard to geographical regions, earlier studies usually focused on areas on the edge of the plateau, including Gansu, Sichuan, Yunnan, Qinghai, East Tibet and areas adjacent to Lhasa in south Tibet. Recent studies have expanded to areas located in central Tibet. However, northwest Tibet, e.g. Kekexili (Hoh Xil) and Ali (Ngari), where the average altitude is 4500m, has still received little attention, which may partially be due to the harsh environment and poor road systems.

Allozymes are among the commonly used markers in early studies, but DNA-based markers (e.g. RAPD, ISSR, AFLP, microsatellite, and DNA sequences) are becoming more and more widely used in the studies of plant genetic diversity. As a highly variable co-dominant marker, microsatellite (i.e. SSR) has a few outstanding advantages in population genetic studies (Selkoe and Toonen 2006).

However, the lack of sequence information has restricted the use of microsatellite in Tibetan plants. Instead, several anonymous markers (e.g. RAPD and ISSR) have frequently been used because of their convenience and low cost. With the advance of sequencing techniques, more microsatellite primers can be developed in the near future. In addition, haplotype data based on chloroplast or mitochondrial DNA sequences have been commonly used in the investigation of the phylogeographic structure of plant species.

AMOUNT AND DISTRIBUTION OF GENETIC VARIATION

To assess the levels and partitioning of genetic variation in plants from the Qinghai-Tibetan Plateau, we have reviewed the papers published in recent years. Only data based on samples from natural populations have been considered, and the studies focusing on crop varieties have been excluded. In addition, some studies based on inadequate sampling strategies (e.g. only 3-4 individuals from each population) were also excluded.

Data for 19 species are summarized in Table 1. Generally, most studies were based on RAPD and/or ISSR markers and most pecies were perennial herbs distributed above the tree line, i.e. alpine plants (*sensu* Körner 1995). As mentioned earlier, these species can be classified into four major categories: 1) wild relatives of crops, e.g. *Elymus sibiricus* L. and *Roegneria thoroldiana* (Oliv.) Keng; 2) endangered species, e.g. *Pinus squamata* X. W. Li and *Pedicularis dunniana* Bonati; 3) medical plants, e.g. *Anisodus tanguticus* (Maximowicz) Pascher, *Lamiophlomis rotata* (Benth.) Kudo, *Swertia przewalskii* Pissjaukova, and several *Rhodiola* species; and 4) ecologically important species (i.e. dominant species in local communities), e.g. *Androsace tapete* Maxim., *Polygonum viviparum* L., and *Sophora moorcroftiana* (Benth.) Baker. Please note that this is not a strict classification and the four categories are not mutually exclusive to each other.

1. Genetic Diversity at Population Level

Because of the unequal numbers of populations of a species investigated in different studies (ranging from 2 to 10), the species level genetic diversity may not be comparable. Accordingly, we only considered intrapopulation variation that is measured using Nei's gene diversity index (H_s) and Shannon's diversity index (I).

(a)

(b)

Figure 2. (a) *Androsace tapete*, a cushion plant, is an important ecosystem engineer species in the alpine ecosystem of the Qinghai-Tibetan Plateau. (b) *Rhodiola crenulata*, a medicinal plant endemic to the Qinghai-Tibetan Plateau. (c) *Lamiophlomis rotata*, a medicinal plant endemic to the Qinghai-Tibetan Plateau.

Previous studies have suggested that the levels of genetic variation within populations are significantly affected by several factors including life form, mating system, population history, and effective population size (Hamrick and Godt 1996, Booy et al. 2000). Here, 18 of the 19 species under investigation are perennial herbs. In most cases, the information about the mating system is not available. Generally, the degree of population genetic variation observed varied greatly between species, Hs ranging from 0.017 to 0.319 with a mean of 0.173. This value is lower than the avarage value of both short-lived (Hs = 0.20) and long-lived perennial plant species (Hs = 0.25) (Nybom 2004). The average lower diversity in plants endemic to the Qinghai-Tibetan Plateau may be partitially due to the unusually low values in two endangered species. Specifically, Zhang et al. (2005) detected an extremely low level of genetic diversity (Hs = 0.017, I = 0.025) in two populations of *Pinus squamata*, which is a highly endangered pine consisting of only 32 individuals. A similarly low value was found in the endangered *Pedicularis dunniana* (Hs = 0.062), with a single population of no more than 50 individuals. In contrast, Geng et al. (2008) found high levels of genetic diversity (Hs = 0.319, I = 0.467) in five populations of *Androsace tapete*. Such high diversity could be attributed to its unique life history characteristics (longevity and long juvenile phase) and large population size (Geng et al. 2008).

Table 1. Genetic diversity of endangered species or wild relatives of crops in Tibet. Hs, Nei's gene diversity index; I, Shannon's diversity index; Gst, Nei's genetic differentiation coefficient, Φst, an analogue of Gst based on AMOVA.

Category	Species (Family)	Geographic region	Altitude range (m)	Marker type	Genetic diversity	Genetic differentiation	Reference
Wild relatives of crops	*Elymus sibiricus* (Poaceae)	Sichuan	3200-3600	ISSR	$Hs = 0.181$	$Gst = 0.33$ $\Phi st = 0.425$	Ma et al. (2008)
	Roegneria thoroldiana (Poaceae)	Qinghai, Tibet	4015-4710	SSR	$Hs = 0.49$	$Gst = 0.23$	Jiang et al. (2005)
Endangered plants	*Pinus squamata* (Pinaceae)	Northeast Yunnan	1900	RAPD ISSR	$Hs = 0.017, I = 0.025$ $Hs = 0.025, I = 0.039$	$\Phi st = 0.011$ $\Phi st = 0.024$	Zhang et al. (2005)
	Pedicularis dunniana (Scrophulariaceae)	Sichuan, Yunnan	Not report	ISSR	$Hs = 0.062, I = 0.099$	$\Phi st = 0.7462$	Xia & Guo (2006)
	Swertia przewalskii (Gentianaceae)	Qinghai	3280-3660	RAPD ISSR	$I = 0.27$ $I = 0.25$	$\Phi st = 0.52$ $\Phi st = 0.56$	Zhang et al. (2007)
	Anisodus tanguticus (Solanaceae)	Qinghai, Sichuan, Southeast Tibet	3200-4100	RAPD	$Hs = 0.195$	$Gst = 0.3505$ $\Phi st = 0.3298$	Zheng et al. (2008)
Medical plants	*Rhodiola cremulata* (Crassulaceae)	Yunnan, South Tibet	3890-5150	ISSR	$I = 0.268$	$\Phi st = 0.474$	Lei et al. (2006)
	R. chrysanthemifolia (Crassulaceae)	Southeast Tibet	3600-4800 (estimated)	RAPD	$I = 0.1351$	$\Phi st = 0.773$	Xia et al. (2007)
	R. alsia (Crassulaceae)	Qinghai, Gansu, East Tibet	3470-4900	ISSR	$I = 0.1369$	$\Phi st = 0.703$	Xia et al. (2005)
	Lamiophlomis rotata (Lamiaceae)	Qinghai, Yunnan, East Tibet	4200-5100	RAPD ISSR	$Hs = 0.166, I = 0.248$ $Hs = 0.166, I = 0.251$	$Gst = 0.430$ $Gst = 0.422$	Liu et al. (2006)

Table 1. (Continued)

Category	Species (Family)	Geographic region	Altitude range (m)	Marker type	Genetic diversity	Genetic differentiation	Reference
Ecological important species	*Sophora moorcroftiana* (Fabaceae)	South Tibet	2947-4100	Isozyme	$Hs = 0.122$	$Fst = 0.199$	Liu et al. (2006)
	Kobresia humilis (Cyperaceae)	Gansu, Sichuan, Qinghai	2800-3820	RAPD	$Hs = 0.2126$ $I = 0.3185$	$Gst = 0.1891$	Zhao et al. (2006)
	K. royleana (Cyperaceae)	Gansu, Qinghai	2750-3860	RAPD	$Hs = 0.2446$ $I = 0.3662$	$Gst = 0.1066$	Zhao et al. (2006)
	K. kansuensis (Cyperaceae)	Gansu, Sichuan	3450-3820	RAPD	$Hs = 0.2266$ $I = 0.3369$	$Gst = 0.1438$	Zhao et al. (2006)
	K. tibetica (Cyperaceae)	Gansu, Sichuan, Qinghai	3230-3550	RAPD	$Hs = 0.2521$ $I = 0.3772$	$Gst = 0.1884$	Zhao et al. (2006)
	K. setchwanensis (Cyperaceae)	Gansu, Sichuan	3140-3600	RAPD	$Hs = 0.1997$ $I = 0.2998$	$Gst = 0.2101$	Zhao et al. (2006)
	Megacodon stylophorus (Gentianaceae)	Yunnan, Sichuan	3300-4000	ISSR	$Hs = 0.0532$ $I = 0.0792$	$Gst = 0.727$	Ge et al. (2005)
	Androsace tepate (Primulaceae)	South Tibet	4830-5010	ISSR	$Hs = 0.3193$ $I = 0.4665$	$Gst = 0.1251$ $\Phi st = 0.1385$	Geng et al. (2008)
	Polygonum viviparum (Polygonaceae)	Gansu	2000-3900	RAPD	$Hs = 0.1227$ $I = 0.1804$	$Gst = 0.5743$ $\Phi st = 0.6659$	Lu et al. (2008)

Note: Some species may fall into two categories, e.g. *Rhodiola* species are both endangered and medical species.

Given the large altitude ranges in the Qinghai-Tibetan Plateau, an interesting question is how genetic diversity changes with altitude. The answer is very important because it provides the information necessary to characterise the evolutionary potential and genetic structure of alpine plants, and can help to predict the responses of vertical vegetation zones to climate change (Kinnison and Hairston 2007).

In a recent review, Ohsawa and Ide (2007) summarized four common patterns of genetic variation change with altitude: 1) "L < M > H", i.e. the lower and higher populations have less diversity than those at intermediate levels; 2) "L < M < H", i.e. the lower populations have less diversity; 3) "L > M > H", i.e. the lower populations have greater diversity; and 4) "L = M = H", i.e. no significant change with altitude. Some published studies amongst those considered here compared explicitly the genetic diversity within populations from different altitudes. For example, Zhao et al. (2006) investigated the genetic diversity of five *Kobresia* species from the eastern Qinghai-Tibetan Plateau and found no significant correlation between diversity and altitude. In contrast, using allozyme markers, Liu et al. (2006) found that the genetic diversity of *Sophora moorcroftiana* increased significantly with altitude in terms of expected heterozygosity (Hs) but not observed heterzygosity (Ho).

For other species which the authors did not perform such statistical analysis, we plotted the diversity value against altitude, based on data in the original papers, and looked for possible non-linear patterns (i.e. L < M > H). Where the data suggested a linear increase or decrease, statistically analyses were performed to examine the possible correlation of variability with altitude. Our results revealed that most species show no significant correlation between diversity and altitude (Figure 3), suggesting the existance of other factors that affect genetic diversity more strongly than altitude.

Another point is that, in most cited studies, populations from different altitudes are collected from areas that are also far apart, often from different mountains. In other words, the difference of genetic diversity may represent the combined effects of both vertical and horizontal gradients.

2. Genetic Differentiation between Populations

Knowledge of genetic structure, i.e. the distribution of diversity within and between populations of a species, is important for the conservation of alpine species because it provides useful insights into how the species may respond to climate changes. For example, if a large proportion of the diversity resides within populations, this would seem good for *in situ* conservation of alpine species for at least two reasons: 1) the local populations may have high evolutionary potential and thus increase their chances to pass through the environmental filter caused by changed selection regimes; and 2) a large proportion of diversity within populations usually means effective gene exchange between populations, which would help the warm-adapted alleles in low altitude populations to spread into higher populations and thus decrease the risk of local extinction by warming.

A commonly used statistical parameter for genetic differentiation is G_{st} (Nei, 1973), which provides a measure of the proportion of the total diversity occurring between populations. The values of G_{st} (or Φ_{st}, an analogue of G_{st} based on AMOVA) for 19 species endemic to the Qinghai-Tibetan Plateau are presented in Table 1. Most species show considerable genetic differentiation between populations, with G_{st} ranging from 0.1066 to 0.727 and Φ_{st} ranging from 0.011 to 0.773.

The mean value (G_{st} = 0.300, Φ_{st} = 0.481) is largely comparable to the average for short-lived perennial plant species (G_{st} = 0.32, Φ_{st} = 0.41) and higher than that for long-lived perennial plants (G_{st} = 0.25, Φ_{st} = 0.19) (Nybom, 2004). Generally, several endangered and/or medical plants including the genera *Rhodiola, Pedicularis, Lamiophlomis, Swertia*, and *Anisodus* show high genetic differentiation between populations, which may be partitially ascribed to their fragmented habitats and shrinking population size because of overexploitation (Liu et al. 2006). An exception is the highly endangered *Pinus squamata*, in which both limited genetic diversity within populations and low genetic differentiation between populations were found as a result of extremely small population size (Zhang et al. 2005). In contrast, several widespread plants like *Androsace tapete and Kobresia* species, whose life histories are similar to those of long-lived trees, have relatively low genetic differentiation between populations (Geng et al. 2008).

Several studies also investigated the genetic differentiation between populations in a spatial context. One of the most widely considered models is isolation-by-distance. In this case the genetic differentiation between populations is predicted to be quantitatively correlated with the corresponding geographic distance (Wright 1943). A Mantel test can be used to examine the correlation between genetic distances and geographic distances (Mantel et al. 2003). For example, Liu et al. (2006) investigated the spatial genetic structure of ten populations of *Sophora moorcroftiana* along the Brahmaputra River (known within Tibet as Yarlung Zangbo River) and reported a significant correlation between genetic and geographic distances (r^2 = 0.50, p = 0.002). Similar findings were reported in *Anisodus tanguticus* (r = 0.345, p = 0.020), *Rhodiola crenulata* (r

= 0.677, p = 0.006), and *Lamiophlomis rotata* (r = 0.688, p = 0.001). In contrast, no significant correlations were found in *Androsace tapete* (r = 0.042, p = 0.446), *Elymus sibiricus* (r = 0.744, p = 0.993), and *Megacodon stylophorus* (r = 0.531, p = 0.146). The lack of significant correlations between genetic and geographical distances in the last three species suggests that the isolation-by-distance model cannot explain the spatial genetic structure of populations of alpine plant species in the Qinghai-Tibetan Plateau. It is notable that some of the assumptions of the isolation-by-distance model may be invalid in the case of Tibetan plateau. Specifically, the model assumes continuous and homogeneous populations, and ignores the effects of habitat characteristics and demographic variation within the range of a species (Slatkin and Maruyama 1975, McRae 2006). Nevertheless, most plants in the Qinghai-Tibetan Plateau occur within a limited range of altitudes, resulting in a belt along land at those altitudes. Given the complex geomorphology and interlaced valleys (i.e. low altitude areas), the distributions of most alpine plants are not spatial continuously. Recently, a more refined model of isolation-by-resistance has been proposed, which may be more useful in explaining the spatial genetic structure of alpine plants in the Tibetan plateau.

3. Spatial Genetic Structure on a Large Scale: Effects of Landscape Barriers

The Qinghai-Tibetan Plateau has a few outstanding features that make it very different from its counterparts elsewhere in the world. Firstly, it occupies an extremely large area (i.e. 2.5 million km^2) and spans considerable latitude and longitude ranges (i.e. N 25°-40°, E 74°-104°). It is not surprising that the plateau possesses highly diverse vegetation types, including not only vertical zones between different altitudes on the same mountain but also horizontal zones crossing different latitudinal or longitudinal areas (see *The Vegetation of Tibet*, Institute of Botany at the Chinese Academy of Sciences, 1988). Alpine plants on the Tibetan plateau may respond to climate warming by both upward migration along the same mountain, and northward migration in some areas (e.g. northern Tibet) where many plants are continuously distributed. Secondly, many major landscape features (e.g. ridges with peaks of more than 5000m and valleys with basins below 4000 m) run west-east across the plateau, and present significant barriers to the northward migration of alpine plants whose habitats are usually constrained between altitudes of 4000 and 5000m. Thus, knowledge of spatial genetic structure at a large scale, especially across major landscape features, is essential for predicting the response of alpine plants to climate change.

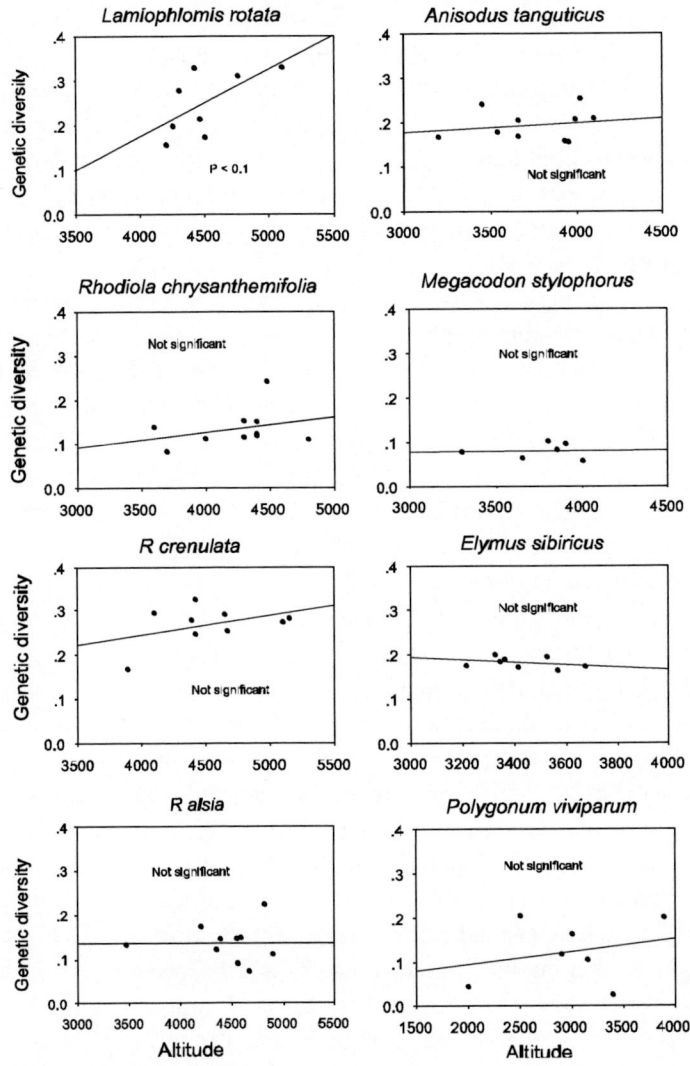

Figure 3. The change of genetic diversity (measured as Hs or I) with altitude in alpine plants endemic to the Qinghai-Tibetan Plateau. The correlation between genetic diversity and altitude is not significant in most species. Only studies including more than six populations are analyzed.

Among the outstanding landscape features of the southern Qinghai-Tibetan Plateau is the Brahmaputra River, which is the largest and longest river on the Tibetan plateau and forms a huge west-east valley, about 1,500 km in length and 200 km in maximum width. Several studies have investigated the effect of the

Brahmaputra River on the spatial genetic structure of alpine plants endemic to Tibet. Using ISSR markers, Xia et al. (2007) investigated the genetic structure of *Rhodiola chrysanthemifolia,* in which five populations were collected to the south and five to the north of the Brahmaputra River. AMOVA revealed that the genetic variation between populations located in the south and north side of the Brahmaputra River was only 4.6%, and most variation (73.1%) was found among populations within regions, suggesting limited genetic differentiation across the Brahmaputra River.

Similarly, Liu et al. (2006) used isozyme markers and analyzed ten populations of the endemic shrub *Sophora moorcroftiana* along the Brahmaputra River. Althought they did not explicitly test the effect of the river on the genetic differentiation among populations of this species. The genetic distances presented in Table 6 of Liu *et al.* (2006) and the results of cluster analysis (Figure 3 in Liu *et al.* 2006) give no indication of genetic discontinuity across the Brahmaputra River. The lack of genetic differentiation across landscape features may result from extensive current and/or past gene exchanges between populations located in different geographical regions. The relative importance of the two factors (current versus historical gene exchanges) can be inferred through the analysis of gene flow at different spatial scales. For example, in a recent study, Geng et al. (2008) explored the spatial genetic structure of *Androsace tapete* at both fine-scale (several meters) and landscape-scale (hundreds of km). On a fine scale, *Androsace tapete* showed significant genetic spatial autocorrelation within a short distance (less than 10 m), suggesting limited current gene dispersal via pollen and/or seeds. On a landscape scale, however, the Brahmaputra River played a weak role in shaping the spatial population structure of this species. The contrasting results of spatial genetic structure at different scales suggest that historical gene exchanges, rather than current gene flow, might have played an important role in shaping the genetic structure of this species across the landscape features like the Brahmaputra River. Besides the Brahmaputra river, a few other landscape barriers (e.g. the Nianqingtanggula and Tanggula mountains) were also involved in the studies of genetic structure in alpine plants endemic to the Qinghai-Tibetan Plateau (Xia et al. 2005). More studies are needed to make a fuller assessment of the effects landscape barriers on the spatial genetic structure of alpine species in the Qinghai-Tibetan Plateau.

CONCLUSION

The population genetics of alpine plants endemic to the Qinghai-Tibetan Plateau have received increasing attention in recent years, as these species are among the most senstive to climate change. Despite significant progress reviewed here, several challenges remain for better management and conservation. Generally, whether or not alpine plants can survive the ongoing climate changes will largely depend on their ability to disperse into suitable new habitats, and on their ability to adapt to the changed environment *in situ* through rapid evolution (Pulido and Berthold 2004). As the dispersal abilities of alpine plants are often difficult to measure using traditional methods, indirect methods of population genetic analysis based on neutral molecular markers represent a promising alternative to assess the gene flow between populations. However there are few studies of this sort. In addition, the knowledge of current vertical and horizontal genetic differentiation can provide important insights into long-term gene exchange and historical dispersal patterns of alpine plants, which are useful for better prediction of their probable responses to future climate change. In addition, efforts are needed to assess the adaptive potentials in alpine plant populations in more accurate ways. Although the neutral molecular markers (e.g. ISSR and RAPD) are most widely used to measure the genetic diversity within natural populations with the assumption of the existance of positively correlationship between marker diversity and the additive genetic variance. However, this assumption has not been tested rigorously, and neutral markers may fail to detect genetic differentiation of great adaptive significance. Thus, there is a need for more well-designed research in Tibet, using both neutral molecular markers and quantitative traits with explicitly ecological significance, to examine the amount and distribution of genetic variation along the altitudinal gradients and across the horizontal zones.

ACKNOWLEDGMENTS

We would like to thank Dr. Tashi Tersing and other colleagues at Tibet University for their help in field investigation and Dr. Yidong Lei, Dr. Jimei Liu, Dr. Qingbiao Wang, Dr. Li Wang and Dr. Liyan Zeng for their assistance in experiments and data analyses. Special thanks to Professors Suhua Shi and Shaoqing Tang for their guidance and support to our studies on plant genetic diversity. This work was supported by Shanghai Science and Technology

Committee (07XD14025), China Postdoctoral Science Foundation (200801171 and 20070410163), and Doctoral Fund of Ministry of Education of China (200802461047).

REFERENCES

Baker, B. B. & Moseley, R. K. (2007). Advancing treeline and retreating glaciers: implications for conservation in Yunnan, PR China. *Arctic Antarctic and Alpine Research*, *39*, 200-209.

Booy, G., Hendriks, R. J. J., Smulders, MJM, Van Groenendael, J. M. & Vosman B. (2000). Genetic diversity and the survival of populations. *Plant Biology*, *2*, 379-395.

Dai, X. K. & Zhang, Q. F. (1989). Genetic diversity of six isozyme loci in cultivated barley of Tibet. *Theoretical and Applied Genetics*, *78*, 281-286.

Ge, X. J., Zhang, L. B., Yuan, Y. M., Hao, G. & Chiang, T. Y. (2005). Strong genetic differentiation of the East-Himalayan *Megacodon stylophorus* (Gentianaceae) detected by Inter-Simple Sequence Repeats (ISSR). *Biodiversity and Conservation*, *14*, 849-861.

Geng, Y. P., Tang, S. Q., Tashi T., Song Z. P., Zhang, G. R., Zeng, L. Y., Zhao, J. Y., Wang, L., Shi, J., Chen, J. K. & Zhong, Y. (2008) Fine- and landscape-scale spatial genetic structure of cushion rockjasmine, *Androsace tapete* (Primulaceae), across southern Qinghai-Tibetan Plateau. Genetica (In press).

Hamrick, J. L. & Godt, M. J. W. (1996). Effects of life history traits on genetic diversity in plant species. *Phil. Trans. Roy. Soc. London Biol. Sci.*, *351*, 1291-1298.

Institute of Botany at the Chinese Academy of Sciences. (1988). *Vegetation of Tibet*. Scientific Press, Beijing.

Jiang, Z. L., Yang, X. M., Wang, R., Gao, A. N. & Li L. H. (2005). Genetic diversity of *Roegneria thoroldiana* (Oliv.) Keng populations based on SSR analyses. *Journal of Plant Genetic Resources*, *6*, 315-318.

Kinnison, M. T. & Hairston, N. G. (2007). Eco-evolutionary conservation biology: contemporary evolution and the dynamics of persistence. *Funtional Ecology*, *21*, 444-454.

Körner, C. (1995). Alpine plant diversity: A global survey and functional interpretations. In Chapin F.S. III and Körner C. (eds), Arctic and Alpine Biodiversity. *Ecological Studies*, *113*, Springer-Verlag, Berlin, 45-62.

Lei, Y. D., Gao, H., Tashi, T., Shi, S. H. & Zhong, Y. (2006). Determination of genetic variation in *Rhodiola crenulata* from the Hengduan Mountains

Region, China using inter-simple sequence repeats. *Genetics and Molecular Biology*, 29, 339-344.
Liu, DS (2000). Implications of fifty year's scientic investigation in Qinghai-Tibetan Plateau. *Resources Science*, 22, 1-5.
Liu, J. M., Wang, L., Geng, Y. P., Wang, Q. B., Luo, L. J. & Zhong, Y. (2006). Genetic diversity and population structure of *Lamiophlomis rotata* (Lamiaceae), an endemic species of Qinghai-Tibet Plateau. *Genetica, 128*, 385-394.
Liu, Z. M., Zhao, A. M., Kang, X. Y., Zhou, S. L. & Lopez-Pujol, J. (2006). Genetic diversity, population structure, and conservation of Sophora moorcroftiana (Fabaceae), a shrub endemic to the Tibetan Plateau. *Plant Biology*, 8, 81-92.
Lu, J. Y., Yang, X. M. & Ma, R. J. (2008) Genetic diversity of clonal plant *Polygonum viviparum* based RAPD in eastern Qinghai-Tibet Plateau of China. *Journal of Northwest Normal University*, 44, 66-72.
Ma, X., Zhang, X. Q., Zhou, Y. H., Bai, S. Q. & Liu, W. (2008). Assessing genetic diversity of *Elymus sibiricus* (Poaceae: Triticeae) populations from Qinghai-Tibet Plateau by ISSR markers. *Biochemical Systematics and Ecology*, 36, 514-522.
McRae, B. H. (2006). Isolation by resistance. *Evolution*, 60, 1551-1561.
Myers, N., Mittermeier, R. A., Mittermeier, C. G., da Fonseca, G. A. B. & Kent, J. (2000). Biodiversity hotspots for conservation priorities. *Nature*, 403, 853-858.
Nei, M. (1973). Analysis of gene diversity in subdivided populations. *Proceedings of the National Academy of Sciences*, USA, 70, 3321-3323.
Nybom, H. (2004). Comparison of different nuclear DNA markers for estimating intraspecific genetic diversity in plants. *Molecular Ecology*, 13, 1143-1155.
Ohsawa, T. & Ide, Y. (2007). Global patterns of genetic variation in plant species along vertical and horizontal gradients on mountains. *Global Ecology and Biogeography*, 17, 156-163.
Pulido, F. & Berthold, P. (2004). Microevolutionary response to climatic change. In: Moller et al (Eds) *Effects of climatic change on birds*. Elsevier, Amsterdam, 151-184.
Qin, D. H. (1998). *The glaciers and ecological environments of the Qinghai-Tibet Plateau*. China Tibetology Publisher, Beijing.
Slatkin, M. & Maruyama, T. (1975). The influence of gene flow on genetic distance. *American Naturalist., 109*, 597-601.

Selkoe, K. A. & Toonen, R. J. (2006). Microsatellites for ecologists: a practical guide to using and evaluating microsatellite markers. *Ecology Letters*, *9*, 615-629.

Wu, C. Y. (1988). Hengduan Mountain flora and her significance. *Journal of Japanese Botany*, *63*, 297-311.

Wu, S. G, Yang, Y. P. & Fei, Y. (1995). On the flora of the alpine region in the Qinghai-Xizang (Tibet) plateau. *Acta Botanica Yunnanica*, *17*, 233-250.

Weng, E. S. & Zhou, G. S. (2006). Modeling distribution changes of vegetation in China under future climate change. *Environmental Modeling and Assessment*, *11*, 45-58.

Xia, J. & Guo Y. H. (2006). ISSR analysis for genetic diversity of *Pedicularis dunniana*. *Journal of Wuhan Botanical Research*, *24*, 565-568.

Xia, T., Chen, S. L., Chen, S. Y. & Ge, X. J. (2005). Genetic variation within and among populations of *Rhodiola alsia* (Crassulaceae) native to the Tibetan Plateau as detected by ISSR markers. *Biochemical Genetics*, *43*, 87-101.

Xia, T., Chen S. L., Chen, S. Y., Zhang, D. F., Zhang, D. J., Gao, Q. B. & Ge, X. J. (2007). ISSR analysis of genetic diversity of the Qinghai-Tibet Plateau endemic *Rhodiola chrysanthemifolia* (Crassulaceae). *Biochemical Systematics and Ecology*, *35*, 209-214.

Xu, W. X. & Liu, X. D. (2007). Response of vegetation in the Qinghai-Tibet Plateau to global warming. *Chinese Geographical Science*, *17*, 151-159.

Zhang, D. F., Chen, S. L., Chen, S. Y., Zhang, D. J. & Gao, Q. B. (2007). Patterns of genetic variation in *Swertia przewalskii*, an endangered endemic species of the Qinghai-Tibet Plateau. *Biochemical Genetics*, *45*, 33-50.

Zhang, Z. Y., Chen, Y. Y. & Li, D. Z. (2005). Detection of low genetic variation in a critically endangered Chinese pine, *Pinus squamata*, using RAPD and ISSR markers. *Biochemical Genetics*, *43*, 239-249.

Zhao, Q. F., Wang, G., Li, Q. X., Ma, S. R., Cui, Y. & Grillo M. (2006). Genetic diversity of five *Kobresia* species along the eastern Qinghai-Tibet Plateau in China. *Hereditas*, *143*, 33-40.

Zheng, W., Wang, L. Y., Meng, L. H. & Liu J. Q. (2008). Genetic variation in the endangered *Anisodus tanguticus* (Solanaceae), an alpine perennial endemic to the Qinghai-Tibetan Plateau. *Genetica*, *132*, 123-129.

Zhou, Z. Q., Shao, Q. Q. & Jiang, X. C. (1984). Comparison of karyotype and chromosome N-banding pattern of *Hordeum spontaneum* of Qing-Zang Plateau and that of the Middle East. *Acta Genetica Sinica*, *11*, 120-124.

INDEX

A

abiotic, 32, 49, 50, 52, 53, 55, 143
absorption, 137, 146
absorption spectroscopy, 146
abundance, 36
acceleration, xiii, 112, 120, 121, 122, 127
accidental, 57
accounting, 56, 58, 62, 98
accuracy, 7
acid, 54, 76, 153
acidity, 82
activation, 65, 119, 122, 125, 126
activation energy, 65
adaptation, 40
adaptive radiation, 177
adult, 27
aerobic, 61
afternoon, 137
age, xiv, 161, 164, 165, 168, 169, 170
aggregation, 15, 31, 35
agriculture, 105
air, ix, xi, 1, 5, 7, 8, 9, 16, 31, 40, 43, 64, 66, 81, 84, 88, 98, 114, 116, 117, 118, 122, 128, 133, 134, 135, 136, 138, 141, 142, 143, 144, 145, 147
algorithm, 126
alleles, 185
alluvial, 124
alps, ix, xi, 80, 82, 83, 99, 100, 102, 106, 107
alternative, 189
alters, 72, 146
amazon, iii
amino, 54, 56, 62, 63, 76
amino acid, 54, 56, 62, 63, 76
ammonium, 54, 85, 87, 101
amoeboid, 71
amplitude, 118
anaerobic, 49, 60, 64, 65, 70
animals, 3, 6, 24, 40, 41, 43, 44, 45, 49, 51, 152, 153
annual rate, 125
anseriformes, 44
Antarctic, 73, 74, 78, 107, 115, 116, 119, 120, 133, 176, 190
anthropic, 176
anthropogenic, xiv, 26, 30, 32, 44, 151, 155, 156
application, 66, 74, 147, 160
aptitude, 14
Arctic Ocean, xi, xii, xiii, 80, 82, 112, 114, 120, 126
arid, 107, 145, 146, 148, 153
army, 84
arrhenius equation, 68

asia, xiv, 44, 46, 151, 155, 156
assessment, 105, 188
assimilation, 63
assumptions, 186
asymptotic, 65
Atlantic, ix, xii, 1, 112, 114, 115, 117, 118, 119, 120, 121, 127, 128, 129
Atlas, 2, 40, 45, 46
atmosphere, xii, xiii, 49, 50, 52, 59, 60, 61, 62, 65, 66, 68, 73, 111, 112, 113, 114, 115, 117, 118, 126, 127, 128, 131, 136, 137, 139, 146, 147, 148, 149
atmospheric pressure, 115, 116, 117, 118, 119, 126
attractiveness, 104
autocorrelation, 188
availability, 82, 100, 103, 105, 106, 108

B

back, xiii, 112, 178
bacteria, x, 48, 50, 51, 55, 56, 57, 58, 62, 67, 68, 71, 72
bacterial, 53, 54, 56, 57, 58, 67, 68, 69, 75, 77
banks, 23
barley, 176, 178, 190
barriers, 70, 186, 188
beaches, 125
behavior, 113
bhutan, 176
bias, 142
biodiversity, xv, 78, 103, 104, 175, 176, 177
biological activity, 83, 101
biological processes, 104
biomass, xiv, 15, 50, 51, 52, 54, 56, 57, 58, 63, 69, 71, 73, 77, 82, 83, 87, 89, 99, 105, 106, 134, 138, 153, 161, 163, 164, 165, 166, 167, 168, 169, 170, 172, 173
biomolecules, 54
biota, ix, x, 1, 6, 43, 47, 49, 52, 69, 70, 71, 78, 152

biotic, ix, x, xiv, 32, 48, 49, 50, 66, 143, 151, 152
biotic factor, 32, 49, 50
birds, 6, 26, 27, 28, 29, 30, 31, 34, 35, 36, 37, 39, 40, 41, 42, 43, 44, 45, 46, 191
Black Sea, 118
body size, 153
bogs, 15, 16, 20, 26, 38
boreal forest, 48, 52, 53, 57, 74, 75
botanical garden, 161
bounds, 32
breeding, 10, 27, 28, 31, 40
Brooks Range, 60, 75
bryophyte, xii, 75, 81, 102
buffalo, 156
buildings, 31, 125

C

calibration, 136
caps, xi, 80, 82
carbon cycling, 132
carbon dioxide, x, 47, 49, 71, 73, 74, 75, 76, 145, 147, 148
carrier, 135
CAS, 145
cast, 12, 13, 15
catabolic, 63
catalyst, 135
catchments, 105, 108
caterpillars, 123
cation, 85, 87
cattle, 34
caucasian, 100
caucasus, 100, 107
cave, 159
CEC, 85, 89, 91, 100
cell, 57, 58, 67, 71
cellulose, 56, 61
census, 45
Central Asia, xiv, 151, 155, 156
Central Europe, 82, 154

Index

changing environment, 71, 82
channels, 114, 123, 124
chemical properties, 88, 89, 92, 102
chemicals, 132
chloroform, 85
chloroplast, 179
chromosome, 178, 192
circulation, xii, 111, 112, 113, 114, 115, 116, 118
classes, 86
classification, 86, 179
clay, 86, 88, 108
climate change, ix, xiv, 1, 7, 28, 45, 72, 74, 76, 77, 82, 109, 113, 114, 116, 117, 121, 132, 133, 151, 152, 153, 155, 156, 158, 159, 162, 184, 186, 189, 192
climate warming, ix, xiii, 1, 26, 28, 76, 112, 114, 121, 122, 126, 127, 152, 186
cluster analysis, 188
clusters, 134, 135
coastal areas, 125
coastal zone, xii, xiii, 112, 114, 121, 122, 123, 126, 127, 128
colonization, 158
combined effect, 184
complement, 132
complexity, 54, 56, 62, 120
compliance, 125
components, xii, 49, 69, 72, 112, 132, 133, 135, 142, 143, 148
composition, xiii, 4, 38, 44, 49, 50, 54, 56, 57, 79, 82, 103, 112, 133, 136, 137, 139, 142, 143, 145, 152, 166, 172, 176
compounds, x, 48, 53, 54, 55, 56, 62, 63, 82, 101
computing, 69
concentration, 49, 55, 66, 82, 87, 89, 95, 99, 100, 126, 135, 137, 138, 139, 141, 143, 144, 145, 147
concrete, 26
conductance, 143, 145
conductivity, 85
congress, vi

conifer, x, 47, 108
coniferous, xi, 80, 82, 101
connectivity, 63
conservation, xv, 44, 45, 101, 152, 176, 177, 184, 189, 190, 191
constraints, 66
construction, 31, 34, 123, 124, 125
construction materials, 125
consumption, 101
control, 103, 104
convection, 126
convective, 114, 118
conversion, 163, 164
cooling, 114, 118, 126
correlation, xv, 55, 115, 176, 184, 185, 186, 187
cotton, 13, 16, 17, 23
coupling, x, 48, 50, 52, 53, 70
covering, xi, 5, 81, 85
creep, 122
critical variables, 52
crops, 5, 176, 178, 179, 182
cross-country, 4
crown, 164
crust, 152
cryogenic, xiii, 12, 112, 122
cultivation, 122, 127
cyanobacteria, 50
cycles, 13, 27, 37, 39, 49, 64, 106, 120, 148, 152, 159
cycling, 44, 54, 59, 105, 109, 132
cysts, 57

D

danger, 122, 127
data set, 86
database, 49, 115
decay, 51, 62, 77
decision trees, 159
decisions, 125
decomposition, xi, 44, 56, 81, 98, 102

deduction, xii, 112
deficiency, 6
deficit, 146
deflation, xiii, 21, 22, 23, 113, 123, 125
deforestation, 106
deformation, 115
degradation, xiii, 34, 54, 75, 112, 122, 123, 127
density, xi, 3, 11, 30, 32, 35, 36, 37, 38, 39, 52, 62, 66, 81, 84, 86, 114, 164
Department of Agriculture, 109
deposition, ix, xi, xiv, 69, 81, 98, 100, 102, 161, 162, 165, 169
deposits, xiii, 13, 51, 112, 122
depressed, 51, 70
depression, 54, 138
desert, 53
desiccation, 58
destruction, 113, 120, 122, 123, 125, 127
detection, 135
detritus, 56
deviation, 9, 29, 37, 132, 143
differentiation, 172, 182, 183, 185, 188, 189, 190
diffusion, 66
diode laser, 146
direct observation, 58
discontinuity, 188
discovery, iii
discrimination, 133, 143, 144, 145
discs, 164
disorder, 10
displacement, 82, 152
distraction, xiii, 112
divergence, 123
diversity, xv, 5, 13, 53, 75, 152, 176, 177, 178, 179, 181, 182, 183, 184, 185, 187, 189, 190, 191, 192
division, 119
DNA, 54, 178, 191
draft, 68
drainage, 26, 59, 147, 163
dry matter, 99

drying, 64, 148
duration, 134

E

ears, 168
earth, 84, 118, 123, 133, 176
Eastern Europe, 44, 46
ecological, ix, xi, 5, 42, 43, 48, 70, 71, 82, 86, 96, 108, 125, 153, 158, 160, 172, 177, 189, 191
ecologists, 176, 192
ecology, vi, x, 5, 43, 45, 47, 49, 53, 62, 152
economic losses, 123, 125, 127
education, 157, 190
electrical conductivity, 85
emission, 51, 64, 73, 75
energy, xiii, 49, 65, 73, 74, 112, 113, 122, 123, 125
enterprise, 122
environment, xi, xii, xiii, 31, 41, 49, 69, 71, 81, 96, 98, 102, 103, 107, 112, 114, 121, 127, 153, 165, 178, 189
environmental change, ix, xiv, 52, 126, 151, 152, 156, 157, 159, 177
environmental characteristics, 59
environmental conditions, 60, 82, 101, 102, 104, 126
environmental control, 146
environmental factors, 52
equilibrium, xii, 111, 113
equilibrium state, 113
erosion, xi, xiii, 12, 81, 98, 100, 101, 102, 112, 121, 122, 123, 124, 125, 126, 127
estimating, 61, 69, 71, 163, 164, 191
Eurasia, ix, 1, 42, 46, 152, 154, 156, 157, 160
Europe, 26, 27, 44, 46, 82, 108, 154, 157, 159, 160
evaporation, 118
evapotranspiration, 64, 142
evolution, 106, 113, 126, 189, 190
exchange rate, 145

exploitation, 12, 38, 123, 155
explosions, 34
exponential functions, 66
exposure, 58, 104, 106, 125
extinction, xiv, 151, 152, 155, 156, 185
extraction, xiii, 43, 85, 105, 113, 122, 125
extrapolation, 153
eyes, 3

F

family, 67, 68
FAO, 107
Far East, 114
fatty acids, 57, 75
fauna, ix, x, xiv, 2, 24, 26, 39, 40, 41, 42, 43, 44, 46, 72, 151, 153, 154, 155, 156, 157
feedback, 82
feeding, 23
Fennoscandian, 44
fertility, 54, 109
fertilization, 54, 75
fertilizer, 77
FID, 135
fishing, 31
flame, 135
flame ionization detector (FID), 135
flavor, 65
flood, xiv, 3, 5, 28, 34, 36, 37, 161, 163, 165, 166, 168, 171
flooring, 34
flora, 46, 96, 97, 177, 192
flora and fauna, 46
flow, xii, 59, 62, 112, 114, 118, 123, 125, 135, 136, 147, 148, 188, 189, 191
flow rate, 135
fluctuations, 27, 35, 39, 66, 113, 128, 152, 154, 156, 157, 158
focusing, 179
food, 15, 35, 39, 54, 56, 58, 64, 73, 105, 153
forbs, 153
forecasting, 125, 128

forest ecosystem, 71, 172
forest management, 173
forests, ix, x, xi, xiv, 3, 13, 47, 48, 52, 53, 57, 77, 80, 82, 99, 101, 108, 153, 154, 161, 163, 164, 165, 166, 170, 172, 173
fossil, 156, 160
fowl, 36
fractionation, 147
fragmentation, 82
freshwater, 71
frost, xi, 55, 73, 81, 98
fumigation, 105
fungal, 57, 69
fungi, x, 48, 51, 56, 57, 58, 71, 72

G

gas, xiii, 12, 14, 17, 31, 32, 39, 43, 65, 66, 70, 126, 131, 135, 146, 148
gas chromatograph, 135
gas exchange, 148
geese, 36, 37, 39
gene, 49, 179, 182, 185, 188, 189, 191
generalizations, 49
generation, 122
genetic diversity, xv, 175, 177, 178, 179, 181, 184, 185, 187, 189, 190, 191, 192
genetics, xv, 175, 189
geography, 41, 43, 61, 107, 178
geology, 103
gigantism, 158
Glacier National Park, 105
glaciers, 106, 120, 190, 191
global climate change, ix, xiv, 151
global warming, vii, viii, x, xii, xiv, xv, 47, 48, 49, 50, 52, 53, 54, 58, 59, 61, 62, 69, 70, 72, 82, 111, 120, 126, 151, 156, 175, 176, 192
glucose, 62, 63
glycine, 77
gonads, 28
GPP, 66

Gram-negative, 57
Gram-positive, 57
grants, 145, 172
graph, 67, 68
grasses, 3, 5, 13, 14, 15, 16, 17, 22, 24, 38, 42, 134, 138, 155
grassland, 60, 72, 97, 133, 145, 147, 148, 154, 155
grazing, 5, 6, 10, 11, 15, 16, 17, 20, 21, 23, 24, 34, 35, 37, 38, 39, 138, 154
greenhouse, xiii, 49, 50, 51, 70, 126, 131
greenhouse gas, xiii, 126, 131
greening, 15
groups, x, 6, 27, 35, 36, 37, 48, 54, 56, 57, 58, 67, 69, 97
growth, 10, 12, 15, 31, 32, 34, 38, 54, 76, 152, 164, 168, 173
guidelines, 173

Holocene, xiv, 40, 49, 106, 123, 124, 151, 153, 154, 155, 156, 157, 159, 160
homogeneity, 123
homogenous, 154
horizon, xi, 81, 84, 85, 86, 90, 98, 99, 100
horses, 153, 155
house, 25, 29, 31
human, 10, 82, 123, 124, 133, 155, 158
human activity, 124
humic substances, 56
humidity, 87, 98, 102
hunting, 26, 155
hydro, 119, 125
hydrodynamic, xiii, 112
hydrological, 70, 76, 120, 127
hydrology, 58, 65, 66, 70, 83, 128
hyperbolic, 66
hypothesis, 102

H

habitat, 28, 82, 152, 158, 176, 186
half-life, 63
haplotype, 179
harvesting, 176
hazards, 125
heat, xii, 6, 112, 114, 117, 118, 122, 126
heating, 58, 60, 118
height, 3, 5, 15, 17, 38, 51, 123, 134, 135, 136, 143, 144, 168
helicopters, 31
hemisphere, 109, 114, 115, 118
herbivores, 52, 152, 153, 154, 155
herbivory, 77
herbs, 5, 138, 179, 181
heterogeneity, 74, 103, 153
heterogeneous, 96
heterotrophic, x, 48, 50, 54, 57, 67
heterozygosity, 184
high pressure, 115
high temperature, 75, 138
high-frequency, 40

I

ice, xi, xiii, 3, 55, 61, 80, 82, 103, 109, 112, 119, 120, 121, 122, 123, 124, 125, 126, 127, 152
ice caps, xi, 80, 82
imagery, 52, 66, 74
images, 5
immigration, 12
immobilization, 106
impact assessment, 105
in situ, 54, 76, 185, 189
independent variable, 64
Indian, xii, 112, 113, 116
Indian Ocean, xii, 112, 113, 116
indication, 188
indicators, 113
indices, 164, 168, 169, 173
indigenous, 10
industrial, 13, 122, 123, 125, 127
industry, 39
infrastructure, 13
inhibition, 57, 71

injury, vi
inorganic, 82, 101
insight, 83
inspection, 27, 65
insulation, 98
interaction, ix, xi, xii, 48, 55, 58, 77, 112, 113, 114
interface, 73, 77
interrelationships, 86
interval, 30, 143, 152
inversion, 128
invertebrates, 54
ionization, 135
IPCC, 132, 147
island, xiv, 31, 35, 121, 123, 125, 161, 163
isolation, xv, 154, 176, 185
isothermal, 101
isotope, 133, 135, 136, 137, 138, 139, 140, 141, 142, 143, 144, 145, 146, 148
isozyme, 178, 188, 190

J

Japanese, 192

K

karyotype, 178, 192
Kazakhstan, 46
kinetics, 56
Kola Peninsula, 26
Kyrgyzstan, 176

L

lagoon, 125
lakes, 5, 23, 73
land, xiv, 3, 5, 21, 23, 28, 34, 39, 41, 82, 90, 99, 105, 161, 163, 165, 166, 168, 186
landscapes, 10, 34, 103, 153, 154, 176
land-use, 105
Lapland, 25, 36, 38, 39, 105

large-scale, 3, 10, 34, 38, 39, 40, 61, 113, 114, 115, 117, 119, 127, 128, 178
laser, 82, 146
Last Glacial Maximum, 152, 155, 160
Late Quaternary, 158, 160
law, 66
leaching, 83, 101, 102
lichen, 3, 5, 16, 17, 18, 20, 23, 24, 38
life forms, 49
lignin, 56, 61
likelihood, x, 48
limitations, 64, 69, 72, 101
linear, 13, 65, 67, 68, 74, 83, 128, 137, 184
linear model, 65, 68
linear regression, 65, 74
links, 35
livestock, 10, 34, 38, 176
living conditions, 40, 126
loading, 11, 12, 15, 38
localised, 12
location, 2, 52, 163
long period, 9, 115
longevity, 181
losses, 59, 123, 125, 127
low molecular weight, 72
low temperatures, xi, 81, 98, 106
lux based, 142
lying, 48

M

magnetic, vi
maintenance, 177
Mammalian, viii, 43, 151, 156, 158, 160
mammals, 40, 41, 154, 157, 158, 160
management, ix, 1, 104, 173, 189
map unit, 55
mapping, 6, 42, 45
marsh, 55
mass loss, 70
mass transfer, 117
matrix, 86, 91, 92

measurement, 60, 113, 135, 138, 142, 143, 145
measures, 125
media, 25, 29
median, 57
medical plant, 176, 178, 179, 185
Mediterranean, 118, 119, 128, 148
melting, 48, 70, 120, 123
meta-analysis, 60, 76
metabolic, 50, 62, 63, 68, 69
metabolism, 50, 56, 61, 69
metabolites, 53
meteorological, x, 6, 7, 8, 48, 52, 64, 104, 113, 119
methane, 70, 108
microbes, x, 48, 50, 51, 54, 56, 58, 63, 64, 67, 69, 75, 76, 101
microbial, ix, x, 48, 50, 51, 52, 53, 54, 56, 57, 58, 60, 61, 62, 63, 67, 68, 69, 71, 72, 73, 76, 77, 79, 82, 85, 87, 89, 90, 96, 101, 105, 106, 107, 108
microbial activity, 105
microbial communities, ix, x, 48, 50, 51, 53, 54, 55, 56, 57, 58, 62, 63, 69, 71, 72, 76
microbiota, vii, x, 47, 48, 49, 50, 57, 58, 66, 68, 69
microclimate, 104
microflora, 73
micrometeorological, 132, 146
microorganism, 69
Middle East, 192
migration, 11, 186
mineralization, 50, 54, 63, 73, 76, 82, 102, 106
mineralized, x, 48, 61
minerals, 108
mines, 12, 13, 15
Ministry of Education, 157, 190
mitochondrial DNA, 179
mixing, 114, 136, 143
modeling, 49, 61, 65
models, ix, xi, 48, 51, 56, 59, 62, 63, 64, 66, 68, 69, 70, 78, 105, 120, 126, 128, 158, 185

moisture, 49, 52, 59, 60, 61, 64, 66, 67, 69, 73, 75, 77, 104, 134, 138, 153, 154
moisture content, 59, 66, 67
molecular markers, 189
molecular weight, 56, 72
molecules, 54
molting, 36
Mongolia, 154, 159
monsoon, 133
morning, 137
morphological, 96
morphology, xii, xiii, 96, 103, 112
mortality, 58, 165
mosaic, 83, 103, 153
mountains, xi, 10, 12, 31, 34, 48, 80, 82, 184, 188, 191
mouth, 27, 171
movement, 11, 26, 27, 28, 41
multidimensional, 86, 107
multidimensional scaling, 107
multivariate, 86, 107

N

National Academy of Sciences, 78, 191
National Science Foundation, 159
natural, xii, 12, 15, 27, 31, 34, 35, 39, 41, 49, 59, 75, 109, 111, 112, 113, 114, 122, 125, 126, 127, 158, 162, 179, 189
natural environment, xii, 31, 49, 112, 114
natural habitats, 34
natural resources, 41
NEC, 39
neck, 48
needles, 166
nematodes, 54
nesting, 24, 28, 29, 30, 31, 32, 34, 35, 36, 37, 40, 42
network, 12, 113
nitrate, 54, 85, 87, 101
nitrification, 102

nitrogen, xii, 44, 50, 53, 54, 55, 70, 73, 75, 76, 77, 81, 82, 83, 85, 87, 95, 96, 98, 100, 101, 105, 106, 107, 108
nitrogen compounds, 53, 55
nitrous oxide, 106, 108
non-random, 115
North America, 116, 152, 156
North Atlantic, 118
Northeast, 109, 182
Northern Hemisphere, 82, 127, 152
novelty, xiv, 151, 155, 156
NPP, 138
nuclear, 191
nutrient cycling, 54, 59
nutrients, ix, xi, 51, 53, 54, 55, 58, 59, 62, 72, 76, 77, 80, 82, 83, 100, 101, 102, 103, 104, 105, 108, 153, 158
nutrition, 53

O

observations, 3, 4, 6, 13, 23, 27, 66, 118, 153
oceans, 114, 115, 119, 126, 128
offshore, 120, 121
oil, 12, 34, 39, 69, 123, 135
oil samples, 135
online, 143, 145
opposition, 114
organ, 153
organic, x, xi, xii, xiv, 48, 49, 50, 51, 53, 54, 55, 56, 62, 63, 69, 70, 71, 72, 77, 81, 82, 83, 85, 87, 96, 98, 99, 100, 102, 106, 121, 127, 132, 134, 137, 138, 143, 147
organic compounds, x, 48, 62, 63, 82
organic matter, x, xi, 48, 49, 50, 56, 63, 64, 69, 70, 77, 81, 82, 83, 98, 99, 100, 102, 106, 137, 143, 147
overexploitation, 185
overgrazing, ix, 1, 15, 18, 20, 34, 35, 176
oxide, 106, 108

P

Pacific, xii, 112, 113, 114, 115, 116, 117, 118, 119, 128, 129
parameter, 185
partition, 133, 142, 145
pastures, 11, 15, 22, 23, 99, 135, 152
peat, xiv, 3, 57, 124, 132, 138
pedestrian, 4
percentile, 85
percolation, 53
periodic, 27, 39, 56
periodicity, 13, 27, 34, 37, 120
permafrost, x, xiii, 3, 48, 49, 51, 56, 69, 73, 106, 109, 112, 121, 122, 126, 152, 153, 168
permit, 66
perturbations, 52
petroleum, 125
pH values, 98
phosphate, 54
phosphorus, 54, 73, 106
photon, 66
photosynthesis, xiii, 50, 51, 131, 132, 136, 137, 138, 141, 142, 143, 145
photosynthetic, 50, 53, 66, 138, 143, 144
phylogenetic, 177
physical properties, xi, 81, 84
physiological, 28, 59, 62, 66, 71
physiology, x, 41, 47
pioneer species, 102
pipelines, 12, 13, 122, 127
planetary, xii, 111, 112, 113
plastic, 23, 65, 135
play, 103, 138
Pleistocene, 46, 124, 152, 153, 154, 155, 156, 157, 158, 160
ploughing, 26
polar bears, 152
polar ice caps, xi, 80, 82
pollen, 154, 159, 160, 188
polygons, 45, 55, 61
polymorphism, 45

pond, 71
pools, 51, 62, 63, 83, 85, 87, 90, 99, 100, 101, 104, 106, 108, 146, 166, 168
poor, 178
population, xv, 3, 12, 27, 28, 30, 31, 34, 40, 41, 43, 44, 45, 56, 59, 67, 68, 86, 102, 107, 122, 127, 152, 156, 175, 177, 178, 179, 181, 185, 188, 189, 191
population growth, 12
population size, 68, 181, 185
pores, 66
porosity, 66
power, 143
precipitation, x, 48, 51, 53, 64, 70, 84, 134, 146, 168, 176
predators, 35, 39, 54, 58
prediction, 70, 189
predictive model, 62
predictors, 60, 64
press, 15, 43, 45, 190
pressure, 16, 24, 45, 115, 116, 117, 118, 119, 126, 135, 146, 156
private, 10, 11
probability, x, 48, 113
probe, 143
producers, 50
production, x, 13, 15, 23, 44, 47, 48, 49, 50, 52, 53, 54, 56, 63, 65, 66, 73, 75, 78, 82, 99, 101, 102, 105, 146, 147, 148, 153, 163, 164, 165, 168, 173
productivity, x, 12, 44, 47, 53, 55, 60, 76, 78, 99, 153, 162, 173
profitability, xiii, 113
program, 145
projector, 34
property, vi
protection, 42, 43, 123, 124, 125
protocols, 85
protozoa, x, 48, 57, 58, 71, 72
protozoan, 57
proxy, 83
pruning, 164
pulse, 56, 78, 101

Q

quantitative estimation, 115

R

radiation, 64, 118, 158, 177
radius, 115
rain, 56
rainfall, 60, 154
random, 113, 115
range, xi, xii, 52, 55, 57, 60, 66, 67, 68, 81, 82, 83, 95, 99, 100, 134, 135, 143, 177, 182, 183, 186
RAPD, 178, 179, 182, 183, 189, 191, 192
RAS, 40, 41, 42, 43, 44, 45, 46, 127, 128
raw material, xiii, 113
real time, 132
reciprocal interactions, ix, xi, 48
reclamation, 125, 127
reconstruction, 7, 29, 42, 43, 154, 160
recovery, 5, 6, 13, 23
recreation, 104
recreational, 12
recurrence, 9, 29, 39
recycling, 54, 153
redistribution, 115, 126
reflection, 26
refractory, 56
refuge, 157
regional, xii, 64, 65, 66, 70, 74, 105, 112, 113, 114
regression, 64, 65, 67, 72, 74
regression analysis, 65
regression equation, 67
regular, 27, 30, 118
regulations, 125
relationship, x, 48, 50, 57, 64, 70, 76, 83, 86, 137, 147, 159, 177
relatives, 176, 178, 179, 182
reproduction, 6, 27, 28, 35
reserves, 23, 69

reservoir, 5, 48, 125
residues, 54
resilience, 156
resistance, 186, 191
resources, 17, 23, 41, 45, 50, 51, 56, 153, 173, 178
respiration, x, xiii, xiv, 48, 51, 54, 57, 59, 60, 61, 62, 63, 64, 66, 67, 68, 69, 73, 75, 76, 78, 132, 133, 136, 137, 138, 141, 142, 143, 145, 146, 147, 148, 165
respiratory, x, 47, 49, 50, 51, 52, 53, 56, 57, 58, 59, 60, 61, 62, 63, 64, 65, 66, 67, 68, 69, 70
restructuring, 153
retention, 100, 101, 103
retrenchment, 52
rhizosphere, 53, 54, 72
ribosomal, 75
rings, 7, 164
risk, 104, 177, 185
rivers, 2, 27
rocky, 34, 153
rodents, 6, 34, 35, 39, 45, 52
roughness, 73
rural, 147
rural areas, 147
Russian, vii, xiii, 1, 40, 41, 42, 43, 44, 45, 46, 111, 112, 127, 128, 157, 161, 172, 173
Russian Academy of Sciences, 1, 161

S

sample, 67, 68, 85, 86, 135, 136, 142, 143, 163, 176
sampling, xi, xii, 60, 67, 81, 84, 85, 95, 107, 135, 143, 179
sand, xi, 21, 23, 81, 85, 86, 88, 91, 123, 124
satellite, 52, 66, 74, 119
satellite imagery, 52, 66, 74
saturation, 49, 51, 85, 87, 100
scaling, 66, 78, 107
sea ice, 122, 125

sea level, xiii, 6, 112, 115, 116, 117, 126
search, 86
seasonal variations, 64
seasonality, 101, 153
secular, 114, 115, 121
sediment, 3, 55, 121, 122, 123, 125, 127, 159, 163
seeds, 188
semi-arid, 145, 146, 148
sensitivity, 72, 132
septum, 148
sequencing, 179
series, 26, 31, 56
services, vi
settlements, 3, 10, 12, 13, 27, 31, 32, 34, 35
severity, 32
shape, 83
shaping, 188
shares, 166
sheep, 135
shelter, 153
shoot, 23
shorebirds, 152, 159
shores, xiii, 112, 119, 121, 122
short period, 27, 28
short-term, 27, 142, 146
shrubs, 3, 5, 13, 15, 16, 17, 22, 23, 24, 39, 134, 138, 153
signals, 146
signs, 7, 23, 113, 114, 119, 122
similarity, 6
simulations, 64, 154
skeleton, 86, 98
skills, 10
small mammals, 40, 154
Smithsonian Institution, 159
smoothness, 3
SOC, 55
soil erosion, xi, 81, 100, 101
solar, 49, 64
solar energy, 49
South America, 116, 118

spatial, xv, 13, 64, 66, 76, 105, 126, 143, 176, 185, 186, 188, 190
specialization, xi, 80, 82
speciation, 158
species richness, 58, 153, 156
specificity, 26
spectroscopy, 146
spectrum, 95, 96, 97
speculation, ix, 1
speed, 119, 126, 143
springs, 9, 28, 29, 31
SPSS, 86
stability, 25, 114, 122, 127
stages, 46, 164, 178
stainless steel, 135
standard deviation, 95, 96
statistical analysis, 184
steady state, 65
steel, 34, 135
stock, 5, 55, 101
storage, 55, 63, 71, 76, 83, 107, 132, 147, 149
storms, 56, 126
strategies, 104, 179
stratification, 113
streams, 2
stress, 86
substances, 23, 56
substrates, 63
sugars, 56
summer, 7, 8, 9, 11, 29, 35, 43, 57, 61, 73, 79, 84, 101, 106, 118, 120, 121, 134, 135, 153, 154
superposition, 113
supervision, 7
supply, 5, 54, 75, 76, 121, 123, 127
surface layer, 126
surface roughness, 73
surface water, 114
survival, 152, 156, 177, 190
survival rate, 152
surviving, 154, 155
synanthropic, 31
systems, 54, 60, 70, 125, 132, 178

T

taiga, 73, 77, 152, 154, 163, 172
tangible, 123, 127
taxa, xii, 58, 67, 68, 81, 102, 152
taxonomic, 67, 160
teflon, 135
temporal, 64, 76, 78, 143, 165
terraces, 23
territory, 3, 11, 12, 13, 14, 15, 23, 32, 34, 39, 41
thawing, x, 48, 122
third order, 120
threat, 125
threatened, 177
thymidine, 54
tides, 123
time, ix, xiii, 1, 7, 10, 17, 20, 23, 26, 27, 28, 29, 30, 31, 35, 39, 56, 63, 97, 99, 102, 112, 113, 114, 115, 118, 119, 120, 121, 123, 125, 127, 132, 135, 138, 139, 141, 143, 165
time periods, 165
timing, 103
TOC, 85, 87, 89, 100
top-down, 54
topographic, 83, 101, 103
topsoil, 85, 86, 87, 92, 95, 96, 99, 100
toxic, 154
trading, 12, 31, 34
traits, 155, 189, 190
transfer, xii, 112, 113, 116, 118
transformation, xii, 26, 28, 34, 41, 56, 64, 101, 104, 112, 115
transgression, 121
transition, ix, xiii, xiv, 35, 46, 101, 108, 132, 143, 154, 159, 160, 161, 162, 163
transmission, 30, 122
transpiration, 65, 172
transport, 12, 13, 66, 122, 123, 127
transportation, 43
traps, 6
travel, 3, 114, 118

trees, xiv, 3, 7, 34, 42, 43, 53, 157, 159, 161, 163, 164, 165, 167, 169, 185
turbulence, 118
turbulent, 113, 118
turnover, 52, 63, 72, 74, 77, 101

U

U.S. Department of Agriculture, 109
uncertainty, 63
Urals, 12, 30, 31, 37, 40, 41, 42, 43, 44, 45, 46, 157, 172
UV radiation, 158

V

values, 34, 35, 39, 56, 63, 65, 67, 68, 86, 87, 89, 96, 97, 98, 99, 100, 115, 137, 170, 181, 185
vapor, 74, 146
variability, xii, 7, 52, 69, 74, 98, 102, 103, 111, 113, 115, 118, 128, 172, 177, 184
variables, x, xii, 48, 49, 50, 52, 55, 61, 62, 64, 66, 81, 86, 98, 99, 102
variance, 66, 67, 86, 97, 189
variation, xv, 52, 55, 59, 60, 66, 67, 71, 73, 74, 97, 118, 119, 128, 143, 146, 148, 164, 175, 177, 179, 181, 184, 186, 188, 189, 190, 191, 192
vehicles, 31
velocity, xiii, 112, 116
vertebrates, ix, x, 2, 146
village, 12, 31
visible, 3
voles, 35, 36, 39
vortex, 113
vulnerability, 104

W

waste products, 51, 63
wastes, 50
water, 12, 28, 36, 49, 51, 53, 59, 60, 61, 64, 65, 66, 70, 72, 73, 74, 75, 82, 89, 92, 103, 114, 120, 121, 122, 123, 127, 138, 148
water table, 51, 53, 60, 61, 64, 65, 66, 70
water vapor, 74
waterfowl, 45
watershed, xiv, 3, 5, 36, 37, 38, 73, 78, 161, 163, 165
weathering, 83, 108
web, 54, 56, 58, 64
Western Europe, 27
Western Siberia, vii, ix, 1, 6, 12, 27, 28, 30, 31, 40, 41, 42, 43, 44, 45, 46, 75, 168
wetlands, 146
wheat, 71
wildlife, vi
wind, xiii, 12, 112, 115, 116, 120, 122
winter, xi, 31, 59, 61, 73, 81, 83, 98, 101, 104, 106, 120, 133, 135, 152, 153, 154
wood, 7, 12

Y

Y-axis, 68
yield, 84, 163

Z

zoology, 41